Grade 8 · Unit 2

Inspire Science

Energy and Motion

Phenomenon: What is growing on these magnets?

This image shows iron filings between two magnets. The image shows how a magnetic field between the magnets affects the iron filings.

Birds use Earth's magnetic field to navigate!

FRONT COVER: Eric Tischler/Alamy Stock Photo. **BACK COVER:** Eric Tischler/Alamy Stock Photo.

mheducation.com/prek-12

Copyright © 2020 McGraw-Hill Education

All rights reserved. No part of this publication may be reproduced or distributed in any form or by any means, or stored in a database or retrieval system, without the prior written consent of McGraw-Hill Education, including, but not limited to, network storage or transmission, or broadcast for distance learning.

Send all inquiries to:
McGraw-Hill Education
STEM Learning Solutions Center
8787 Orion Place
Columbus, OH 43240

ISBN: 978-0-07-687490-3
MHID: 0-07-687490-7

Printed in the United States of America.

7 8 9 LMN 24 23 22 21 20

McGraw-Hill is committed to providing instructional materials in Science, Technology, Engineering, and Mathematics (STEM) that give all students a solid foundation, one that prepares them for college and careers in the 21st century.

Welcome to
Inspire Science
Explore Our Phenomenal World

Learning begins with curiosity. Inspire Science is designed to spark your interest and empower you to ask more questions, think more critically, and generate innovative ideas.

Start exploring now!

Inspire Curiosity • Inspire Investigation • Inspire Innovation

Authors, Contributors, and Partners

Program Authors

Alton L. Biggs
Biggs Educational Consulting
Commerce, TX

Ralph M. Feather, Jr., PhD
Professor of Educational Studies and Secondary Education
Bloomsburg University
Bloomsburg, PA

Douglas Fisher, PhD
Professor of Teacher Education
San Diego State University
San Diego, CA

Page Keeley, MEd
Author, Consultant, Inventor of Page Keeley Science Probes
Maine Mathematics and Science Alliance
Augusta, ME

Michael Manga, PhD
Professor
University of California, Berkeley
Berkeley, CA

Edward P. Ortleb
Science/Safety Consultant
St. Louis, MO

Dinah Zike, MEd
Author, Consultant, Inventor of Foldables®
Dinah Zike Academy, Dinah-Might Adventures, LP
San Antonio, TX

Advisors

Phil Lafontaine
NGSS Education Consultant
Folsom, CA

Donna Markey
NBCT, Vista Unified School District
Vista, CA

Julie Olson
NGSS Consultant
Mitchell Senior High/Second Chance High School
Mitchell, SD

Content Consultants

Chris Anderson
STEM Coach and Engineering Consultant
Cinnaminson, NJ

Emily Miller
EL Consultant
Madison, WI

Key Partners

American Museum of Natural History

The American Museum of Natural History is one of the world's preeminent scientific and cultural institutions. Founded in 1869, the Museum has advanced its global mission to discover, interpret, and disseminate information about human cultures, the natural world, and the universe through a wide-ranging program of scientific research, education, and exhibition.

PhET Interactive Simulations

The PhET Interactive Simulations project at the University of Colorado Boulder provides teachers and students with interactive science and math simulations. Based on extensive education research, PhET simulations engage students through an intuitive, game-like environment where students learn through exploration and discovery.

SpongeLab Interactives

SpongeLab Interactives is a learning technology company that inspires learning and engagement by creating gamified environments that encourage students to interact with digital learning experiences. Students participate in inquiry activities and problem-solving to explore a variety of topics through the use of games, interactives, and video while teachers take advantage of formative, summative, or performance-based assessment information that is gathered through the learning management system.

Measured Progress, a not-for-profit organization, is a pioneer in authentic, standards-based assessments. Included with New York Inspire Science is **Measured Progress STEM Gauge**® assessment content which enables teacher to monitor progress toward learning NGSS.

Table of Contents
Energy and Motion

Module 1 Forces and Motion

Encounter the Phenomenon 3
STEM Module Project Launch 4

Lesson 1 Position and Motion 5
   Science Probe Train Ride 5
   Encounter the Phenomenon 7
   Explain the Phenomenon Claim/Evidence/Reasoning Chart 8
   Investigation Follow the Directions 10
   Investigation Start from Here 11
   Investigation See you soon 13
   LAB Watch it go 15
   LAB Be the Fastest 20
   How It Works GPS to the Rescue! 24
   Investigation Point the Way 25
   Investigation Plot It 27
   Review 30

Lesson 2 Force and Acceleration 33
   Science Probe Constant Mowing 33
   Encounter the Phenomenon 35
   Explain the Phenomenon Claim/Evidence/Reasoning Chart 36
   LAB Up to Speed 38
   Investigation When Push Comes to Shove 41
   LAB Sticky Situation 45
   Investigation Diagram a Force 48
   LAB A Balancing Act 50
   STEM Careers A Day in the Life of a Vehicle Crash Test Engineer 53
   Review 54

Lesson 3 Force Pairs 57
   Science Probe Blowing in the Wind 57
   Encounter the Phenomenon 59
   Explain the Phenomenon Claim/Evidence/Reasoning Chart 60
   LAB Pulling Your Weight 62
   Investigation Back to Back 65
   LAB Bounce Back 68
   A Closer Look SAFER Barriers 71
   Review 72

Lesson 4 Gravitational Force 75
   Science Probe Ball Toss 75
   Encounter the Phenomenon 77
   Explain the Phenomenon Claim/Evidence/Reasoning Chart 78
   LAB Use the Forces 80
   Investigation The Pencil Dropped Around the World 81
   Investigation The Force of Gravity 83
   LAB Weighing Washers 87
   Investigation Gravity of Objects 90
   A Closer Look Space Travel 91
   Review 92

STEM Module Project Engineering Challenge: Crash Course 95

Module Wrap-Up 103

Module 2 Mechanical Energy

Encounter the Phenomenon... 105

STEM Module Project Launch 106

Lesson 1 Kinetic Energy 107
   Science Probe Soccer Ball 107
   Encounter the Phenomenon 109
   Explain the Phenomenon Claim/Evidence/Reasoning Chart 110
   Investigation Rolling On 112
   LAB Mass Matters 113
   LAB Picking Up Speed 117
   A Closer Look Kinetic Energy 121
   Review 122

Lesson 2 Potential Energy 125
   Science Probe Don't Fall 125
   Encounter the Phenomenon 127
   Explain the Phenomenon Claim/Evidence/Reasoning Chart 128
   LAB Slingshot Physics 130
   Investigation Dropping the Ball 133
   STEM Careers A Day in the Life of a Roller Coaster Designer 137
   Review 138

Lesson 3 Conservation of Energy 141
   Science Probe Swing Low 141

Table of Contents (continued)
Energy and Motion

Encounter the Phenomenon 143
Explain the Phenomenon Claim/Evidence/Reasoning Chart 144
LAB The Energy of a Pendulum 146
LAB So Much Work 151
LAB Double Pendulum 154
A Closer Look Creating Electrical Energy 157
Review 158

STEM Module Project Science Challenge: Energy at the Amusement Park 161

Module Wrap-Up 167

Module 3 Electromagnetic Forces

Encounter the Phenomenon 169

STEM Module Project Launch 170

Lesson 1 Magnetic Forces 171
Science Probe Which pole is it? 171
Encounter the Phenomenon 173
Explain the Phenomenon Claim/Evidence/Reasoning Chart 174
LAB Paper Clip Pick Up 176
LAB The Strength of Magnets 178
LAB Magnetic Personality 180
LAB Magnetic Fields 182
LAB Moving Magnets 187
LAB Create a Magnet 190
A Closer Look Magnetic Migration 193
Review 194

Lesson 2 Electric Forces 197
Science Probe Electric Charge 197
Encounter the Phenomenon 199
Explain the Phenomenon Claim/Evidence/Reasoning Chart 200
LAB From Top to Bottom 202
LAB Paper Pick Up 205
Investigation Field Rings 208
How It Works Van de Graaff Generator 213
Review 214

Lesson 3 Simple Circuits 217

Science Probe Plugging In 217
Encounter the Phenomenon 219
Explain the Phenomenon Claim/Evidence/Reasoning Chart 220
LAB Lighten Up 222
LAB Power Up 225
Science & Society A Smart Grid? 229
Review 230

Lesson 4 Electromagnetism 233
   Science Probe Charged Magnets 233
   Encounter the Phenomenon 235
   Explain the Phenomenon Claim/Evidence/Reasoning Chart 236
   LAB Pointing Directions 238
   Investigation Making Magnetic Fields 240
   LAB Electromagnet Challenge 243
   LAB Motor On 247
   Engineering Investigation Electric Motor Mechanics 248
   LAB Coiled Up 250
   Investigation Lights On 252
   LAB Lift it up 254
   STEM Careers A Day in the Life of a Maglev Train Engineer 257
   Review 258

STEM Module Project Engineering Challenge: The Great Metal Pick Up Machine 261
Module Wrap-Up 267

Forces and Motion

ENCOUNTER
THE PHENOMENON

Why is the boat still moving if no one is rowing?

Row Your Boat

GO ONLINE
Watch the video *Row Your Boat* to see this phenomenon in action.

Collaborate With your class, develop a list of questions that you could investigate to find out more about the motion of the boat. Record your questions below.

Module: Forces and Motion

STEM Module Project Launch
Engineering Challenge

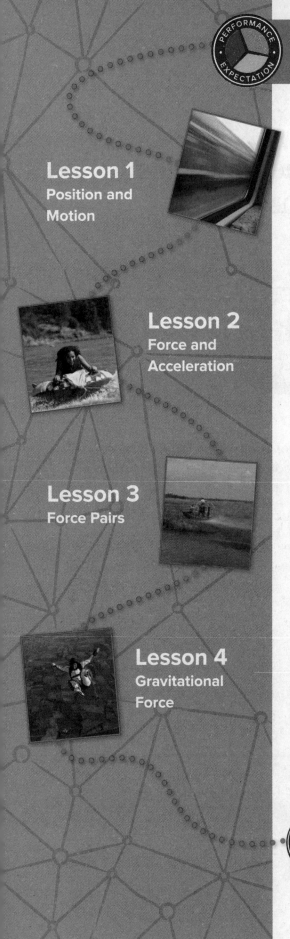

Lesson 1 Position and Motion

Lesson 2 Force and Acceleration

Lesson 3 Force Pairs

Lesson 4 Gravitational Force

Crash Course

Like the rowers at the beginning of the module, a car will continue to move unless acted on by a force. If the brakes fail while moving, the car will remain in motion. The mayor of a nearby town has hired your group of engineers to develop a solution to help protect drivers and reduce the damage that might result from a car or other vehicle striking a new bridge support.

The mayor would like your group to design, test, and evaluate a possible solution to the problem. In addition, the mayor needs you to provide results of investigations to show that the change in motion of a vehicle depends on the sum of forces and mass to help inform other city leaders of the science behind the solution.

Start Thinking About It

The image above shows one possible solution to the town's problem. How do you think this solution protects drivers and the bridge support?

STEM Module Project
Planning and Completing the Engineering Challenge How will you meet this goal? The concepts you will learn throughout this module will help you plan and complete the Engineering Challenge. Just follow the prompts at the end of each lesson!

LESSON 1 LAUNCH

Train Ride

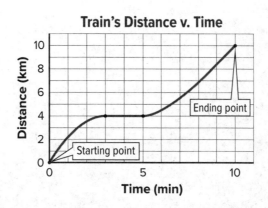

The graph shows a train traveling between two stations. Circle what you think is the best interpretation of the graph.

A. The train sped up during the first 3 minutes, slowed down for several minutes, then sped up again.

B. The train went uphill, traveled on flat ground, then went uphill again.

C. The train sped up for 3 minutes, traveled at a steady speed for 2 minutes, then sped up again.

D. The train sped up then slowed down during the first 3 minutes, stopped for two minutes, then sped up again.

E. The train traveled at a steady speed uphill for most of the way except for in the middle when it stopped for 2 minutes.

Explain the answer you selected to describe the motion of the train.

_____
_____
_____
_____
_____
_____

You will revisit your response to the Science Probe at the end of the lesson.

LESSON 1
Position and Motion

ENCOUNTER THE PHENOMENON

How can you describe the position and motion of the train outside the window?

Imagine you are sitting on a train. Using the photo on the previous page for visual clues, describe the position and motion of the train outside the window.

Now describe the position and motion of the train you are sitting on.

GO ONLINE
Watch the video *Get Moving!* to see this phenomenon in action.

ENGAGE Lesson 1 Position and Motion 7

EXPLAIN THE PHENOMENON

When you described the position and motion of the trains, what type of words did you use? Did you use *fast* or *slow*? Did you use *near* or *far*? What about the words *right* or *left*? Is one description better than another to know the position and motion? Make a claim about the most accurate way to describe a train's position and motion.

CLAIM
The position and motion of the train is best described by...

 COLLECT EVIDENCE as you work through the lesson. Then return to these pages to record your evidence.

EVIDENCE

A. What evidence have you discovered to explain how position can be used to describe the train?

B. What evidence have you discovered to explain what motion is?

8 Module: Forces and Motion

MORE EVIDENCE

C. What evidence have you discovered to explain how to describe changes in motion?

D. What evidence have you discovered to explain how a graph can help you understand motion?

When you are finished with the lesson, review your evidence. If necessary, based on the evidence, revise your claim.

REVISED CLAIM

The position and motion of the train is best described by...

Finally, explain your reasoning for how and why your evidence supports your claim.

REASONING

The evidence I collected supports my claim because...

Where are you right now?

Imagine you are boarding a train. How would you describe where you are? You might say you are sitting one meter to the left of your friend. Perhaps you would explain that you are at the train station, which is a half mile north of your school. You might instead say that the station is ten blocks east of the center of town, or even 150 million kilometers from the Sun. The descriptions are useful if they can be repeated.

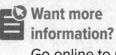

Want more information? Go online to read more about position and motion.

INVESTIGATION

Follow the Directions

How would you give instructions to a friend who was trying to walk from one place to another? Use your classroom for this investigation.

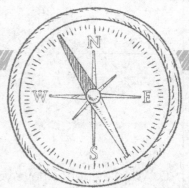

1. Place a sheet of paper labeled *North*, *East*, *South*, and *West* on the floor. This is your starting location.

2. Walk from the paper to one of the three locations your teacher has labeled in the classroom. Have a partner record the number of steps and the directions of movement in the space below.

3. Using these measurements, write instructions other students could follow to move from the starting location to the new location.

4. Repeat steps 2 and 3 for the other locations. Record your instructions in your Science Notebook.

5. Compare your instructions with instructions from another group. How did your instructions to each location compare to those written by other groups?

_____

_____

_____

6. How did the description of your movement depend on the point at which you started?

_____

_____

_____

Describing Position To describe position you must state your location relative to a certain point. A **reference point** is the starting point you choose to describe the location, or position, of an object. The chosen point is arbitrary, which means that you pick the point based on your preference. The reference points from the paragraph on the previous page are your friend, your school, the center of town, and the Sun.

A description of your location also includes your distance and direction from the reference point. Describing your location in this way defines your position. A **position** is an object's distance and direction from a reference point. A complete description of your position includes a distance, a direction, and a reference point. Let's explore the relationships between those three things.

> **FOLDABLES**
> Go to the Foldables® library to make a Foldable® that will help you take notes while reading this lesson.

INVESTIGATION

Start from Here

1. Put a sticky note at the 50-cm mark of a meterstick. This is your reference point.

2. Place a small object at the 40-cm mark. It is 10 cm in the negative direction from the reference point.

EXPLORE/EXPLAIN Lesson 1 Position and Motion **11**

3. Continue moving the object and recording its distance, its reference direction, and its position to complete the table below.

Position of Object		
Distance (cm)	Reference Direction	Position (cm)
10 cm	negative	40 cm
40 cm	positive	
15 cm	positive	
	positive	75 cm
		30 cm

4. How would the data in the table change if the positions were the same, but the reference point was at the 40-cm mark? Complete the table below with the new information.

Position of Object		
Distance (cm)	Reference Direction	Position (cm)

The Reference Direction When you describe an object's position, you compare its location to a reference direction. The reference direction is the positive (+) direction. In the first part of the Investigation *Start From Here*, any number greater than 50 cm was in the positive direction. The opposite direction is the negative (−) direction. As with the reference positon, reference direction is arbitrary. You decide which direction is positive and which is negative. Suppose you specify east as the reference direction in the figure below. You could say the museum's entrance is +80 m from the bus stop. The library's entrance is −40 m from the bus stop.

12 EXPLORE/EXPLAIN Module: Forces and Motion

Moving in Two Dimensions You now know that position can be described with a reference point, distance, and direction. So far you've only experienced this in one dimension—a straight line. Let's explore how to describe position in two dimensions.

INVESTIGATION

See You Soon

Hinano and Marko agree to meet at the state fair. Hinano arrives at the entrance. Marko gives Hinano two-part instructions explaining how to find him. His instructions use three reference points (including her starting point), two directions, and two approximate distances. The instructions describe the shortest walking distance between them. Study the diagram and then answer the questions.

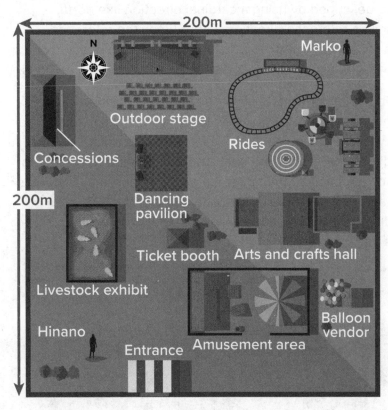

1. What instructions do you think Marko gave Hinano?

___

___

___

___

EXPLORE/EXPLAIN Lesson 1 Position and Motion

2. What is an alternative route for Hinano to find Marko?

_____

_____

Describing Position in Two Dimensions When Marko told Hinano his location using two directions, he was using two dimensions. Marko used meters in his description. Meters are the base unit of length per the International System of Units (SI). Like the reference point and the reference direction, the units used are arbitrary. Scientists use SI units to eliminate confusion of multiple measurement systems.

Position can also be described by using a cardinal direction like *north*, *south*, *east*, or *west* as a reference direction. Other times the reference direction can be as simple as *right or forward*. To describe the position of a window on a building, you might choose *left* and *up* as reference directions.

Finding a position in two dimensions is similar to finding a position in one dimension. First, choose a reference point. To locate your classmate's home on the map below, you could use your home as a reference point. Next, specify reference directions—south and east. Then, determine the distance along each reference direction. In the image below, your classmate's house is one block south and one block east of your house. In order to communicate the directions to a friend, you have to share the position, direction, and units that you are using.

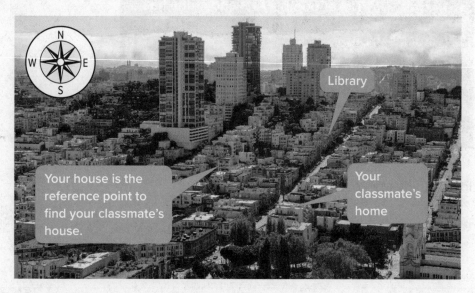

COLLECT EVIDENCE

How can you describe the position of the train from the beginning of the lesson? Record your evidence (A) in the chart at the beginning of the lesson.

14 EXPLORE/EXPLAIN Module: Forces and Motion

What is motion?

Think back to the train at the beginning of the lesson. If the train is moving from north to south, do you have a complete understanding of the train's motion? If the train is traveling south for 460 km, now do you have a complete understanding of the train's motion? The answer to both questions is no. There are additional factors to consider. Let's find out what they are.

OPEN INQUIRY LAB Watch It Go

Safety

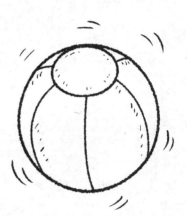

Materials

meterstick
tennis ball
masking tape
stopwatch
paper towel tubes
large rubber band
wind-up toys

Procedure

1. Read and complete a lab safety form.

2. Write a set of procedures on the next page that you will use to measure and describe an object's motion. You may or may not use all the materials listed above. Include in your procedure:

    A. The materials you will use.

    B. The method you will use to make the object move.

    C. How you will keep the object moving in a straight line.

    D. How you will measure the distance and time, including the units you will use.

    E. The type of data table you will need to record your data.

    F. The number of trials you will complete.

Procedure, continued

3. Have your teacher approve your procedures. Follow your approved procedures to complete your investigation. Record your observations and data on the next page.

4. After you complete your investigation, follow your teacher's instructions for proper cleanup.

Data and Observations

Analyze and Conclude

6. Identify the independent and dependent variables, as well as the control(s), in your investigation.

_____
_____
_____

7. Describe the motion of the object you tested. Include as many directions and measurements in your description as you can.

_____
_____
_____
_____

Motion When you described the motion of the object in the Lab *Watch It Go*, you should have included how the object's position changed. You can tell the object moved because its position changed relative to the reference point. **Motion** is the process of changing position.

Observing Motion Look below at the figure of the man fishing. Is the man in motion between the top part of the figure and the bottom part? Suppose the fishing pole is the reference point. Because the positions of the man and the pole do not change relative to each other, the man does not move relative to the pole. Now suppose the buoy is the reference point. Because the man's distance from the buoy changes, he is in motion relative to the buoy. All motion is relative to a certain other point in space.

Motion Using Reference Points Suppose you are watching a soccer game like the one in the figure below. The position of a player depends on a reference. If the reference point is the goal, or point A, the player's position is 10 m in front of the goal. If the reference point is center field, point B, the position of the player is 40 m toward the goal. Notice that the actual location of the player has not changed. Only the description of the position changed because the reference point changed.

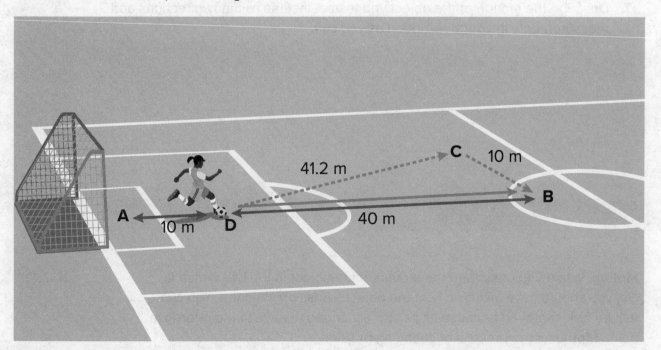

18 EXPLORE/EXPLAIN Module: Forces and Motion

Distance and Displacement During one play in the soccer game on the previous page, the player runs 41.2 m from position D to position C. Then she runs 10 m to position B. Her path is shown by the green dotted lines. The total distance the player travels is 41.2 m + 10 m = 51.2 m.

The solid purple arrow in the figure shows the player's displacement. **Displacement** is the difference between the initial, or starting, position and the final position. The player starts at point D and finishes at point B. Her displacement is 40 m in front of her initial position. Displacement is the shortest distance between where the player started and the player's final position. An object's displacement and the distance it travels are not always equal. If the player runs directly from point D to point A, then both the player's distance and displacement are the same quantity—10 m. If the player's final position is the same as her starting position, her displacement is 0 m.

THREE-DIMENSIONAL THINKING
Use the race track **model** below to determine the distance traveled and the displacement of a car from point A to when it reached point D on the first lap.

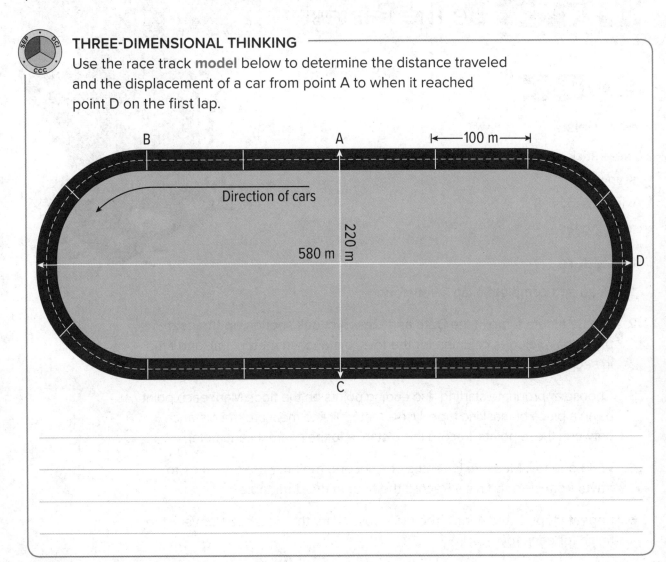

COLLECT EVIDENCE
How can you describe the motion of the trains from the beginning of the lesson? Record your evidence (B) in the chart at the beginning of the lesson.

EXPLORE/EXPLAIN Lesson 1 Position and Motion

What do you measure to determine motion?

How fast do you walk when you are hungry and there is good food on the table? How fast do you move when you have a chore to do? Sometimes you move quickly, and sometimes you move slowly. Let's find out how to measure the motion.

LAB Be the Fastest

Safety

Materials

meterstick
stopwatch
wind-up toys (4)
masking tape

Procedure

1. Read and complete a lab safety form.

2. Create a data table in the Data and Observations section on the next page that includes columns for the toys, distance in meters (m), and time in seconds (s).

3. Choose appropriate starting and ending points on the floor. Mark each point using a piece of masking tape. Use a meterstick to measure the distance between these points. Record this distance to the nearest centimeter.

4. Wind one toy. Measure in tenths of a second the time the toy takes to travel from start to finish. Record the time in the data table.

5. Repeat steps 3 and 4 for three more toys. Vary the distance from start to finish for each toy.

6. Follow your teacher's instructions for proper cleanup.

Data and Observations

Analyze and Conclude

7. Using the data you collected, when was the motion of the toys changing and when was the motion constant?

   _____
   _____
   _____
   _____

8. How could you use the data you collected to describe the motion of the wind-up toys?

   _____
   _____
   _____
   _____

Changes Over Time In the Lab *Be the Fastest*, did all of the toys move at the same rate? Did some move slowly and others move quickly? Describing how fast something moves is the same as determining its speed. **Speed** is a measure of the distance an object travels in a given amount of time.

Constant and Changing Speed Speed can be constant or changing. Look at the figure below. The stopwatches above the girl show her motion every second for 6 seconds. In the first 4 seconds, the girl moves with constant, or unchanging, speed because she travels the same distance during each second. When the girl starts running, the distance she travels each second gets larger and larger. The girl's speed changes because she is moving a different distance each second.

Average Speed The speed of most moving objects is not constant, which is why the speedometer in a car is always changing slightly. Therefore, when you describe your speed over an entire trip to someone, you are describing average speed. Average speed is equal to the total distance traveled divided by the total time. Average speed can be modeled mathematically using the equation below.

Average Speed Equation

$$\text{average speed (in m/s)} = \frac{\text{total distance (in m)}}{\text{total time (in s)}}$$

$$\bar{v} = \frac{d}{t}$$

The symbol $\bar{v}$ represents the term "average velocity." You will read more about velocity, and how it relates to speed, later in the lesson. At this point $\bar{v}$ is simply used as the symbol for "average speed." Imagine the girl above traveled 25 m from second 1 to second 7 on the stopwatch. Therefore, her average speed is 25 m/6.0 s or 4.2 m/s.

MATH Connection The motion of a person or object can be explained by examining how the position changes over time. Practice using the mathematical model, the average speed equation.

1. A truck driver makes a trip that covers 2,380 km in 28 hours. What is the driver's average speed in km/h?

2. What is the average speed of a soccer ball that travels 34 m in 2.0 s?

3. How long would it take a bus traveling at 52 km/h to travel 130 km?

THREE-DIMENSIONAL THINKING
Isaiah leaves one city at noon. He has to be at another city 186 km away at 3:00 PM. The speed limit the entire way is 65 km/h. Can he arrive at the second city on time? Explain your reasoning using **mathematical evidence**.

EXPLORE/EXPLAIN Lesson 1 Position and Motion

HOW IT WORKS

GPS to the Rescue!

You've seen the signs tacked to streetlights and telephone poles: *LOST! Golden retriever. Reward. Please Call!* Losing a pet can be heartbreaking. Fortunately, there's an alternative to posting fliers—a pet collar with a Global Positioning System (GPS) chip that helps locate the pet. Here is how GPS can help you track or locate your pet:

1 GPS is a network of at least 24 satellites in orbit around Earth. Each satellite circles Earth twice a day and sends information to ground receivers.

2 GPS satellites act as reference points. Ground-based GPS receivers compare the time a signal is transmitted by a satellite to the time it is received on Earth. The difference indicates the satellite's distance. Signals from as many as four satellites are used to pinpoint a user's exact position.

3 GPS uses computer technology to calculate location, speed, direction, and time. The same GPS technology used to locate or guide airplanes, cars, and campers can help find a lost pet anywhere on Earth!

4 A GPS pet collar works much the same as any other GPS receiver. Once it is activated, the collar can transmit a message to a Web site or to the owner's cell phone.

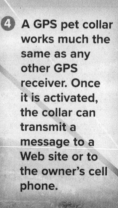

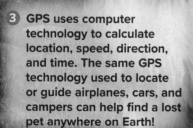

It's Your Turn

READING Connection Research other applications of GPS. What industries use GPS, and how do those industries benefit? After completing your research, build a slideshow presentation. Be sure to use several sources and include any questions that your research generated.

INVESTIGATION

Point the Way

Have you ever been asked for directions and pointed to where the person needed to go? You were indicating direction. Similar to pointing, motion can be represented with arrows. An arrow can point left, right, up, or down indicating direction. An arrow's length can vary to indicate speed. A longer arrow would represent a faster speed.

1. An airplane flew from San Francisco to Washington, D.C. Approximately halfway through the flight, the plane had traveled 2,000 km in 2.5 hours. What was the speed during this period?

2. Another airplane is flying in the opposite direction. It covers the same distance in exactly 2 hours. What was its speed and direction during this period?

3. Draw arrows representing the speed and direction of the two planes. Label each arrow with the speed and direction of flight. Use a left-facing arrow for *west* and a right-facing arrow for *east*.

First plane	Second plane

EXPLORE/EXPLAIN Lesson 1 Position and Motion

Speed and Direction In the Investigation *Point the Way,* you identified the speed and direction of two planes. Both speed and direction are part of motion. **Velocity** is the speed and direction of a moving object. Velocity is referred to as a vector. A **vector** is a quantity that has both magnitude and direction. To share the direction and magnitude, arrows are used to represent vectors. The plane traveling to Washington, D.C. was represented with a smaller vector than the plane flying into San Francisco.

Constant and Changing Velocity Like speed, velocity can either be constant or changing. When velocity is constant, the object is moving at a constant speed and its direction does not change. If either the speed or the direction of the object changes, the velocity will change.

Look at the figure below. It is a motion diagram. A motion diagram is a series of images showing the positions of moving objects at equal time intervals. A motion diagram can be simplified by replacing the objects with dots. Velocity is shown by adding arrows to the dots. Each dot represents a time frame such as 1 s. Notice how from one position to the next, the arrows showing the velocity of the cyclists change length. The arrows can also show a change in direction. The changes in the arrows mean that the velocity is constantly changing.

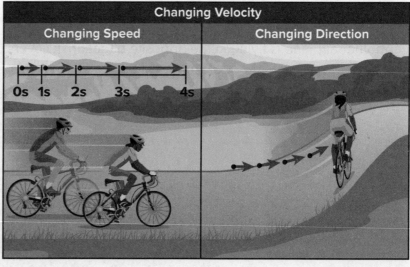

COLLECT EVIDENCE
What factors would you measure to describe changes in the motion of the trains from the beginning of the lesson? Record your evidence (C) in the chart at the beginning of the lesson.

How can a graph help you understand an object's motion?

When you study motion, you need to know how position changes as time passes. Graphs are one type of model that can show how one measurement compares to another. Two measurements frequently compared to each other are distance and time. Let's investigate these relationships graphically.

26 EXPLORE/EXPLAIN Module: Forces and Motion

INVESTIGATION

Plot It

1. Plot the distance and time data for distance the train traveled on the grid below. Plot the distance on the vertical axis and time on the horizontal axis. Label the axes and add a title to your plot.

Distance Traveled by Train	
Time (h)	Distance (km)
0	0
1	110
2	220
3	330
4	400
5	500

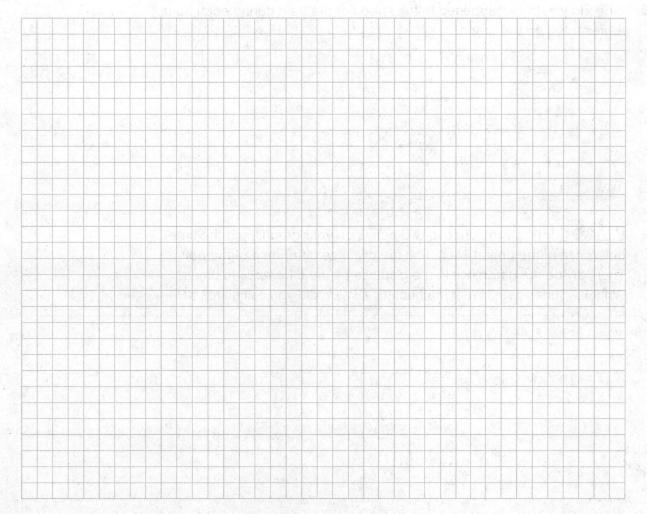

2. Draw a line that goes through the points on the grid in order.

EXPLORE/EXPLAIN Lesson 1 Position and Motion 27

3. What does the graph tell you about the distance traveled by the train?

_____

_____

_____

_____

4. What does the graph tell you about the amount of time that has passed?

_____

_____

_____

_____

5. What do you think happened to the speed of the train during each hour represented on the graph?

_____

_____

_____

_____

6. **MATH Connection** Choose two points on your graph. Find the time difference of the points. Next, find the distance difference of the points. Then, divide the difference in the distance by the difference in time. What is your answer? What mathematical value did you calculate?

Distance-Time Graphs The plot you made in the Investigation *Plot It* is a distance-time graph. This type of graph shows how an object's position changes during each time interval. A distance-time graph does not show you the actual path the object took.

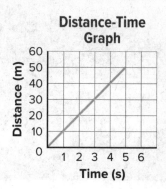

Did you notice that the line through the points in the investigation wasn't completely straight? When an object is moving at a constant speed, the line will be straight. The steeper the line, the greater the slope, which means the greater the speed of the object.

If the slope of the line changes, this means the speed of the object has changed. Even if the speed has changed, the average speed can still be calculated.

GO ONLINE for additional opportunities to explore!

Investigate position and motion by performing one of the following activities.

☐ **Use models** in the **PhET Interactive Simulation** *The Moving Man*.

OR

☐ Find **patterns** by examining changes over time in the **Lab** *Calculate Average Speed from a Graph*.

THREE-DIMENSIONAL THINKING

Analyze the data on the plot below. Determine the speed of the hawksbill sea turtle during each interval listed below.

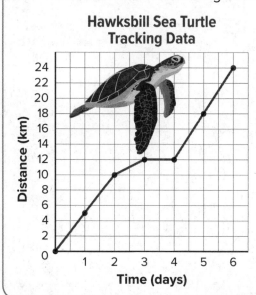

Day 0 to day 2: _____

Day 2 to day 3: _____

Day 3 to day 4: _____

Day 4 to day 6: _____

COLLECT EVIDENCE

How do time-distance graphs help you understand the motion of the train? Record your evidence (D) in the chart at the beginning of the lesson.

EXPLORE/EXPLAIN Lesson 1 Position and Motion **29**

LESSON 1
Review

Summarize It!

1. **Construct and Present Arguments** A distance-time graph shows the motion of two bicycle riders. Each rider's motion is represented on the graph by a diagonal line sloping upward from left to right. The graph shows that they traveled the same distance. However, the line representing the motion of Rider #1 slopes upward more steeply than the line representing the motion of Rider #2. Sketch a graphical model of the motions of the riders. Develop an argument on which rider arrived at his or her destination first. How do you know? Use evidence to add validity to your argument.

Three-Dimensional Thinking

Analyze the data table below. Use the table to answer questions 2 and 3.

Green Sea Turtle's Distance and Time Data

Time (days)	Distance (km)
0	0
1	16
2	32
3	48
4	64
5	80
6	96

2. The data in the table above shows how far a sea turtle travels over several days. What would the line on a plot of this data look like?

   A The line would curve upward and to the right.

   B The line would go up and down.

   C The line would point straight upward to the right.

   D The line would point upward then downward.

3. If the turtle continued the motion recorded in the data table above, what would his distance be at ten days?

EVALUATE Lesson 1 Position and Motion

Real-World Connection

4. **Interpret Data** The plot below shows the motion of an elevator. Explain its motion.

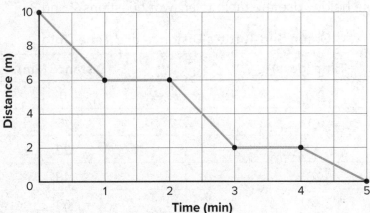

_____

_____

_____

5. **Calculate** A driver travels 55 km in 1 hour. He then drives at a speed of 35 km/h for 2 hours. Next, he drives 175 km in 3 hours. What was his average speed?

_____

Still have questions?
Go online to check your understanding about position and motion.

REVISIT

PAGE KEELEY SCIENCE PROBES

Do you still agree with the statement you chose at the beginning of the lesson? Return to the Science Probe at the beginning of the lesson. Explain why you agree or disagree with that statement now.

EXPLAIN THE PHENOMENON

Revisit your claim about how you can describe position and motion of a train. Review the evidence you collected. Explain how your evidence supports your claim.

START PLANNING
STEM Module Project Engineering Challenge

Now that you understand the differences between position and motion, go to your Module Project to determine the criteria and constraints of your design. Keep in mind any societal or environmental impacts.

LESSON 2 LAUNCH

Constant Mowing

Stacey sees a person pushing a lawn mower across level ground. She is walking at constant speed and pushing the mower with constant force. Could the lawn mower move faster than the person is walking?

- **A.** Yes, but only if she is not walking fast.
- **B.** Yes, but only if she pushed harder and harder.
- **C.** Yes, but only if the force of the push on the mower was greater than the force of friction with the grass.
- **D.** No, because if she was pushing with constant force, the mower would have to move at constant speed.

Which statement do you most agree with? Explain your reasoning.

_____
_____
_____
_____

You will revisit your response to the Science Probe at the end of the lesson.

LESSON 2
Force and Acceleration

ENCOUNTER THE PHENOMENON

What happens to the motion of the water tube when it's pushed or pulled?

GO ONLINE
Watch the video *Pulled Away!* to see this phenomenon in action.

While watching the video, record your observations and the ways you see the motion of objects changing. Think about what causes the change in motion of the water tube.

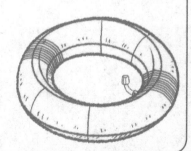

ENGAGE Lesson 2 Force and Acceleration 35

EXPLAIN
THE PHENOMENON

Motion is changing all the time. When you stand up, begin to run, or turn a corner, your motion is changing. These changes in motion all have something in common. They all start with a push or a pull. What happens when you push or pull an object? Make a claim about how a push or pull affects the motion of a water tube.

CLAIM
When you push or pull a water tube...

 COLLECT EVIDENCE as you work through the lesson. Then return to these pages to record your evidence.

EVIDENCE
A. What evidence have you discovered to explain why motion changes?

B. What evidence have you discovered to explain how friction changes motion?

MORE EVIDENCE

C. What evidence have you discovered to explain how multiple forces will change motion?

When you are finished with the lesson, review your evidence. If necessary, based on the evidence, revise your claim.

REVISED CLAIM

When you push or pull a water tube...

Finally, explain your reasoning for how and why your evidence supports your claim.

REASONING

The evidence I collected supports my claim because...

Lesson 2 Force and Acceleration

What can cause a change of motion?

Think about kicking a moving soccer ball. You change the ball's motion, and the velocity changes. With the water tube, it sometimes sped up or slowed down depending on how the boat pulled it. What other kinds of changes in motion can there be? Let's investigate!

Want more information?
Go online to read more about changing motion and forces.

FOLDABLES
Go to the Foldables® library to make a Foldable® that will help you take notes while reading this lesson.

LAB Up to Speed

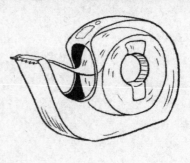

Safety

Materials

masking tape   metronome
meterstick   stopwatch

Procedure – Part I

1. Read and complete a lab safety form.

2. Use masking tape to mark a course on the floor. Mark a starting place, and place marks along a straight path at 10 cm, 40 cm, 90 cm, 160 cm, and 250 cm from the start.

3. Turn on the metronome.

4. On the first beat, the person walking the course is at start. On the second beat, the walker should be at the 10-cm mark, and so on. Record your observations below.

Data and Observations

Analyze and Conclude

5. Explain what happened to your speed as you moved along the course.

___

6. Suppose your speed at the final mark was 0.95 m/s. Calculate your average rate of change from the start through the final segment of the course.

___

Procedure — Part II

7. Now use the meterstick to measure a 6 m straight path along the floor. Place a mark with masking tape at 0 m, 3 m, and 6 m.

8. Look at the graph below. Decide what type of motion occurs during each 5-second period.

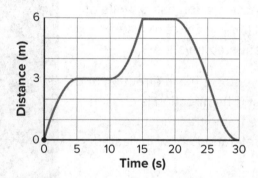

9. Try to walk along your path according to the motion shown on the graph. Have another student time your walk with a stopwatch. Switch roles and repeat this step. Record your observations below.

10. Follow your teacher's instructions for proper cleanup.

Data and Observations

Analyze and Conclude

11. What does a horizontal line segment on a distance-time graph indicate?

_____

_____

12. According to the graph, at what times do the following motions take place?

You change direction. _____

Your speed increases. _____

Your speed decreases. _____

Changes in Motion An object's velocity changes if either the speed or the direction of an object changes. When an object's velocity changes, the object is accelerating. **Acceleration** is a measure of the change in velocity during a period of time. An object accelerates when its velocity changes as a result of increasing speed, decreasing speed, or changing direction. Acceleration has SI units of meters per second per second (m/s/s). This can also be written as meters per second squared (m/s^2).

Like velocity, acceleration is a vector. It has a direction and can be represented by an arrow. The direction is chosen based on an arbitrary reference frame that must be shared. The length of each acceleration arrow indicates the amount of acceleration.

Now that you know that an acceleration is a change in motion, think about what might affect how an object accelerates. What evidence would you need to collect to make a connection about what factors affect an object's acceleration? Record your notes in the first column. Pair with a partner and discuss his or her notes. Write any new notes in the second column. Then record in the third column what you both would like to share with the class.

Think	Pair	Share

40 EXPLORE/EXPLAIN Module: Forces and Motion

The water tube at the beginning of the lesson accelerated as the boat pulled it across the water. Think about what caused the motion to change. What could accelerate the water tube faster? Let's investigate!

INVESTIGATION

When Push Comes to Shove

GO ONLINE Explore the PhET interactive simulation *Forces and Motion: Basics*.

After exploring the simulation on your own, return to the simulation's Home screen and follow the instructions below.

1. From the Home screen click the *Motion* icon, then check the boxes for *Values, Masses,* and *Speed* at the top right.

2. Apply 50 Newtons of force until the speed reaches approximately 20 m/s.

3. Remove the applied force. Describe the motion of the box.

_____
_____
_____
_____

4. Reset the simulation, and check the boxes for *Values, Masses,* and *Speed*.

5. Apply 500 Newtons of force until the speed reaches approximately 20 m/s. Then, remove the force. Describe the motion of the box.

_____
_____
_____
_____
_____
_____

EXPLORE/EXPLAIN Lesson 2 Force and Acceleration

6. Try to stop the box. Explain the steps needed to stop the box.

7. Reset the simulation, and check the boxes for *Values*, *Masses*, and *Speed*. Then, add the refrigerator to the skateboard.

8. Apply 50 Newtons of force until the speed reaches approximately 20 m/s then remove the force. Describe the motion of the refrigerator.

9. The wrapped present on the bottom right has an unknown mass. Write the steps to an investigation that could determine the mass of the present.

10. Once the force was removed, what happened to the speed of the skateboard?

11. What effect did the larger mass of the refrigerator have on the acceleration?

42 EXPLORE/EXPLAIN Module: Forces and Motion

Forces What have you pushed or pulled today? You might have pushed open the classroom door or pulled the zipper on your backpack. A **force** is a push or a pull on an object. There are two ways a force can affect an object. A force can change an object's speed. It also can change the direction in which the object is moving. For a plane to make a turn, a force is required to change the motion in the direction of that turn. In other words, a force can cause acceleration. All acceleration is due to forces. In the Lab *Up to Speed*, your feet were pushing on the floor. They were using force to accelerate you across the floor. Because acceleration is a vector with a direction, force also is a vector with a magnitude and direction.

In the PhET interactive simulation *Forces and Motion: Basics*, a mannequin pushed the skateboard to accelerate it. Think back to the water tube. A boat pulled the rope which pulled the water tube. In each case, an object applied a force to an object that it touched. A **contact force** is a push or a pull on one object by another object that is touching it.

EARTH SCIENCE Connection Contact forces can be weak, like when you press the keys on a computer keyboard. They also can be strong, such as when large sections of underground rock suddenly move, resulting in an earthquake. The large sections of Earth's crust called plates also apply strong contact forces against each other. Over long periods of time, these forces can create mountain ranges if one plate pushes another plate upward.

THREE-DIMENSIONAL THINKING
Make an **argument** that the **cause** of changing the motion of an object is a force. Support your argument with evidence.

_____
_____
_____
_____

EXPLORE/EXPLAIN Lesson 2 Force and Acceleration

Mathematical Model Isaac Newton, an English scientist and mathematician, developed laws of motion. **Newton's second law of motion** states that the acceleration of an object equals the net force on the object divided by the object's mass. With a mathematical model, scientists can make predictions about how objects will accelerate with a given force. This can be shown by the mathematical model, the acceleration equation.

Acceleration Equation

acceleration (in m/s^2) = $\dfrac{\text{force (in N)}}{\text{mass (in kg)}}$

$$a = \dfrac{F}{m}$$

MATH ▶ Connection In the PhET interactive simulation, a force of 100 N is applied to the wrapped present, giving it an acceleration of 2 m/s^2. What is the mass of the object?

What is the acceleration when a force of 2.0 N is applied to a ball that has a mass of 0.60 kg?

COLLECT EVIDENCE

How do the changes in motion help explain what happens when you push or pull a water tube? Record your evidence (A) in the chart at the beginning of the lesson.

How does friction affect motion?

Think about what would happen if you slid a book across a table. Would it continue to slide, or would it come to a stop after some time? What about the water tube? Only a force can change the motion of an object. What forces might be responsible for slowing things down when the push stops?

LAB Sticky Situation

Safety

Materials

tape
sandpaper
eyehook

spring scale
wooden block

Procedure

1. Read and complete a lab safety form.

2. Use tape to fasten sandpaper to a table. Attach a spring scale to a wooden block with an eyehook in it.

3. Gently pull the block at a constant speed across the table. Record your observations below.

4. Gently pull the block at a constant speed across the sandpaper. Record your observations.

5. Follow your teacher's instructions for proper cleanup.

Data and Observations

EXPLORE/EXPLAIN Lesson 2 Force and Acceleration

Analyze and Conclude

6. Compare the forces required to pull the block across the two surfaces.

7. How did changing the surface affect the motion of the block?

8. Which surface created a larger opposing force? Explain.

Friction Rub your hands together. What do you feel? The resistance you feel is friction. **Friction** is a force that resists the sliding motion of two surfaces that are touching. If your hands were soapy, you could slide them past each other easily. You feel more friction when you rub your dry hands together than when you rub your soapy hands together. To the right is a close-up view of two metal surfaces. Microscopic dips and bumps like the ones shown cover all surfaces. When surfaces slide past each other, the dips and bumps on one surface catch on the dips and bumps on the other surface. This microscopic roughness slows sliding and is a source of friction. When you switched to sandpaper in the Lab *Sticky Situation*, the dips and bumps became exaggerated which made the block harder to pull across the surface.

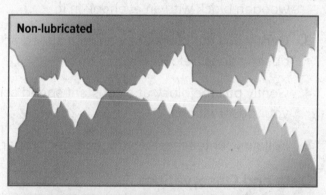

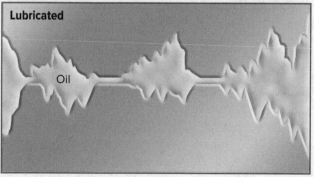

46 EXPLORE/EXPLAIN Module: Forces and Motion

Friction and Newton's Laws Think about how friction affects an object's movement. Imagine a book sitting on a table. When you push the book, the force you apply to the book is greater than the friction between the book and the table. The book moves in the direction of the greater force. If you stop pushing, friction stops the book, just like the friction between tires and the road stops a car.

What would happen if there were no friction between the book and the table? According to Newton's second law, the book would continue to move at the same speed in the same direction because no force changed its motion. The book stopping is evidence that friction must be acting on the book. On Earth, friction can be reduced but not totally removed. For an object to start moving, a force greater than friction must be applied to it. This is why all objects on Earth come to a stop after some time. To keep the object in motion, a force at least as strong as friction must be continuously applied. Objects stop moving because friction or another force acts on them.

THREE-DIMENSIONAL THINKING

If you **cause** the surfaces between two touching objects to change, how will the friction be affected? Make a claim. Support your **explanation** with evidence and logical connections.

_____
_____
_____
_____

COLLECT EVIDENCE

How does friction's effect on motion help explain what happens when you push or pull a water tube? Record your evidence (B) in the chart at the beginning of the lesson.

EXPLORE/EXPLAIN Lesson 2 Force and Acceleration 47

How do multiple forces change motion?

Force has both size and direction. Just as you can use arrows to show the size and the direction of velocity and acceleration, arrows can show the size and direction of a force. It is very common that more than one force is acting on an object at a time. Sometimes motion can change in multiple directions. Scientists often use models called free-body diagrams to understand these changes in motion. A **free-body diagram** is a simple model to understand systems of objects with any amount of applied forces. How are these models used?

INVESTIGATION

Diagram a Force

A circle is drawn to represent the object as shown to the right. An *m* is placed inside to represent the mass of the object. Next, an arrow is drawn to represent the direction that the force is applied. The size of the arrow indicates how strong the push or pull is.

1. Sketch a free-body diagram of an object being pushed to the right.

To model an object with multiple forces, simply add one arrow to the circle in the direction of each force as shown below.

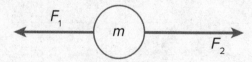

2. In the free-body diagram above, the force to the left (F_1) is less than the force to the right (F_2). Which direction do you think the object will begin to move?

_____

_____

48 EXPLORE/EXPLAIN Module: Forces and Motion

When all the arrows are added together, the result it is called the net force. The **net force** is the sum of all the forces acting on an object. To model the net force, simply add together the forces in the same direction and subtract the forces in the opposite direction. The sizes of the arrows show how much is added or removed. Then draw a free-body diagram of the net force.

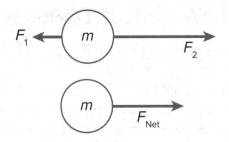

3. Make a free-body diagram for the net force on the object shown below. The object is sliding across a surface with friction.

4. An object is pushed to the right with 10 Newtons of force, and pulled to the left with 20 Newtons. Sketch a free-body diagram of this system, and draw a diagram with the net force. Ignoring friction, identify the direction and motion of the object.

When multiple forces act on an object, the forces will add together and act as a force in one direction. In many cases, the change in motion is in one direction but is a result of many forces. The motion of an object is determined by the sum of the forces acting on it. If the total force on the object is not zero, its motion will change.

EXPLORE/EXPLAIN Lesson 2 Force and Acceleration

How can forces act on an object that is not changing its motion?

Imagine that the water tube from the beginning of the lesson came across some still water, and the boat continued at a constant velocity. The water tube would move at a constant velocity. Would the rope be applying a force on the water tube? Each change in motion is a result of a force, but do all forces change an object's motion?

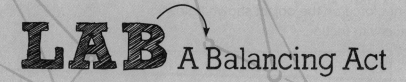 A Balancing Act

Safety

Materials

spring scales (2) wooden block
eyehooks (2)

Procedure

1. Read and complete a lab safety form.

2. Attach spring scales to opposite sides of a wooden block with eyehooks, and place the block and spring scales on a flat surface.

3. With a partner, gently pull the scales so that the block moves toward one of you at a constant speed. Sketch the setup in the Data and Observations section below, including the force readings on each scale.

4. Repeat step 3, pulling on the block so that it does not move. Record your observations.

5. Follow your teacher's instructions for proper cleanup.

Data and Observations

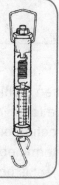

50 EXPLORE/EXPLAIN Module: Forces and Motion

Analyze and Conclude

6. Explain what occurred in steps 3 and 4.

7. What would happen if the block was moving faster in step 3?

8. Sketch a free-body diagram for the forces acting on the block when it was not moving.

9. Under what conditions can an object have forces acting on it, but its motion does not change?

Newton's First Law Recall that Newton's second law of motion states that the acceleration of an object equals the net force on the object divided by the object's mass. If the force acting on an object is zero, then the acceleration must also be zero. This is Newton's first law of motion. **Newton's first law of motion** states that an object in motion will stay in motion, and an object at rest will stay at rest unless acted on by a force.

Balanced and Unbalanced Forces In the figure to the right, students are playing tug of war. Both sides are pulling with the same force, so neither side is winning. If the forces acting on an object are balanced, the object's motion does not change. Balanced forces are forces that cancel each other. The only way for a team to win is to have unbalanced forces, when one team pulls harder than the other team. When the forces are unbalanced, the rope and everyone holding it will accelerate in the direction of the net force.

Balanced forces can act on objects that are moving as well. Recall that for an object to change its motion, a net force must be acting on it. As a boat pulls the water tube at a constant speed, the net force must be zero. According to Newton's first law of motion, balanced forces cause no change in an object's velocity. This is true when an object is at rest or in motion. The skiers in the figure to the right are being pulled up the mountain at a constant velocity. Even when they are moving and pulling on the rope, the net force is zero.

THREE-DIMENSIONAL THINKING

In what ways does Newton's second law of motion describe **stability and change** of any **system** that has forces acting on it?

_____
_____
_____
_____
_____
_____
_____

COLLECT EVIDENCE

How do multiple forces help explain what happens when you push or pull a water tube? Record your evidence (C) in the chart at the beginning of the lesson.

STEM Careers

A Day in the Life of a Vehicle Crash Test Engineer

Engineers work to make crashes safer for the driver and passengers. The airbags and seatbelts in vehicles are specially engineered using Newton's laws. Should a vehicle suddenly stop, the people inside would continue their motion unless acted on by a force.

Engineers know that Newton's first law explains the motion of the crash-test dummy. Before the crash, the car and dummy moved with constant velocity. If no other force had acted on them, the car and dummy would have continued moving with uniform velocity. The impact with the barrier results in an unbalanced force on the car, and the car stops. The seatbelt and airbag create the forces that change the dummy's motion.

It's Your Turn

Research and Report Researchers and engineers use Newton's first and second laws when they are searching for solutions to crashes. In what ways are modern vehicles still taking advantage of Newton's laws? Develop a slide show presentation on how these laws of motion are used in modern cars.

ELABORATE Lesson 2 Force and Acceleration

LESSON 2
Review

Summarize It!

1. **Organize** Create a graphic organizer that compares and contrasts forces, acceleration, friction, balanced forces, and unbalanced forces.

Three-Dimensional Thinking

An object has a force acting on it to the right and has a frictional force to the left as shown below. Use the model below to answer questions 2 and 3.

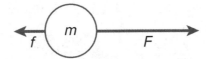

2. What change in motion will result from the forces modeled?

    A There will be no change in motion because the forces are in opposite directions.

    B The object will slow down because of the friction force.

    C The object will accelerate to the right.

    D The object will accelerate to the left.

3. What would a model of the net force look like?

    A The arrow would be to the right at the same length as before, because friction is a different force.

    B The arrow would be to the right but shorter than before to account for the friction force.

    C There would be no net force because the two forces are in opposite directions.

    D The arrow would be to the left because friction is slowing the object down.

---

4. A train moves at a constant speed down a straight track. Which of the following scientific explanations is true?

    A No forces act on the train as it moves.

    B The train moves because no forces are acting against it.

    C The forces of the train's engine balance the force of friction opposing it.

    D An unbalanced force keeps the train moving.

EVALUATE Lesson 2 Force and Acceleration

Real-World Connection

5. **Infer** Your friend is trying to move a couch. No matter how hard he pushes, the couch will not budge. Explain to your friend why the couch is not moving even though he is applying a force.

_____
_____
_____
_____
_____

6. **Explain** A motorboat is accelerating across the water when the engine fails. Explain what will happen to the motion of the boat with no force coming from the engine.

_____
_____
_____
_____

 Still have questions?
Go online to check your understanding about changing motion and forces.

 **REVISIT SCIENCE PROBES** Do you still agree with the statement you chose at the beginning of the lesson? Return to the Science Probe at the beginning of the lesson. Explain why you agree or disagree with that statement now.

 **EXPLAIN THE PHENOMENON** Revisit your claim about how pushes and pulls affect the motion of the water tube. Review the evidence you collected. Explain how your evidence supports your claim.

KEEP PLANNING
STEM Module Project
Engineering Challenge

Now that you've learned about Newton's first and second laws of motion, go back to your Module Project to explain what factors affect acceleration of the vehicle. Keep in mind how Newton's second law of motion will affect your technology.

LESSON 3 LAUNCH

Blowing in the Wind

Paula is sailing across a lake. As the boat accelerates, she thinks about how the wind is pushing on the sail. Which statement below best represents what is happening between the sail and the air?

A. The sail is pushing the wind back and slowing down the air.
B. The sail and air are not pushing each other, they are just moving together.
C. The air is pushing on the sail, but the sail is not pushing the air.
D. The sail is pushing the wind forward with the air.

Which statement do you most agree with? Explain your reasoning.

_____
_____
_____
_____
_____

You will revisit your response to the Science Probe at the end of the lesson.

LESSON 3
Force Pairs

ENCOUNTER THE PHENOMENON | How does the air push the airboat forward?

With a partner, take turns sitting on a skateboard and try pushing your partner away. Record your observations in the space below.

Now sit on a skateboard and push against a wall. Record your observations below.

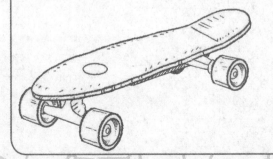

GO ONLINE
Watch the video *Blowing Air* to see this phenomenon in action.

ENGAGE Lesson 3 Force Pairs **59**

EXPLAIN THE PHENOMENON

In the activity, you may have noticed that your motion changed when you were the one that pushed on your partner. When the airboat pushed the air behind it, the boat moved forward. How can pushing something away result in an opposite push or motion? Make a claim about what is happening when the airboat pushes against air.

CLAIM
When an airboat pushes on the air...

COLLECT EVIDENCE as you work through the lesson. Then return to these pages to record your evidence.

EVIDENCE

A. What evidence have you discovered to explain the forces present when you push an object?

B. What evidence have you discovered to explain how to show Newton's third law?

60 Module: Forces and Motion

MORE EVIDENCE

C. What evidence have you discovered to explain the forces during a collision?

When you are finished with the lesson, review your evidence. If necessary, based on the evidence, revise your claim.

REVISED CLAIM

When an airboat pushes on the air...

Finally, explain your reasoning for how and why your evidence supports your claim.

REASONING

The evidence I collected supports my claim because...

Lesson 3 Force Pairs

What forces are present when you push on an object?

If you think about forces you encounter every day, you might notice forces that occur in pairs. For example, if you drop a tennis ball, the falling ball pushes against the floor. The ball bounces because the floor pushes with an opposite force against the ball. How do these opposite forces compare?

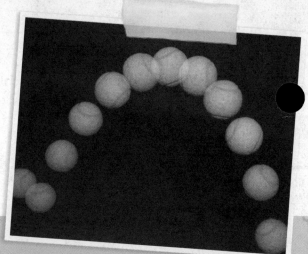

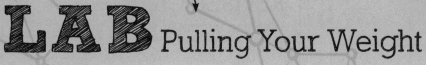

LAB Pulling Your Weight

Safety

Materials

spring scales (2)   key ring

Procedure

1. Read and complete a lab safety form.

2. Stand so that you face a partner, about half a meter away. Each of you should hold your own spring scale.

3. Hook the two scales together with the key ring, and gently pull them away from each other. Record the force reading on each scale below.

4. Pull harder on one of the scales, and again record the force readings on the scales below.

5. Continue to pull on both scales, but let the scales slowly move toward your lab partner and then toward you at a constant speed. Record your observations below.

6. Follow your teacher's instructions for proper cleanup.

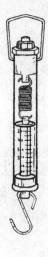

Data and Observations

62   EXPLORE/EXPLAIN Module: Forces and Motion

Analyze and Conclude

7. What patterns did you observe about the directions of the forces from the scales?

___

___

___

8. What relationship did you notice between the force readings on the two scales?

___

___

___

9. If you were to change the direction of your pull, what do you think would happen to the force from your partner's scale?

___

___

___

10. Sketch a free-body diagram of the key ring.

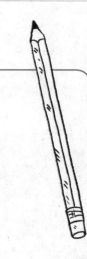

EXPLORE/EXPLAIN Lesson 3 Force Pairs

Opposing Forces In the activity at the beginning of the lesson, you moved because the wall exerted a force on you. This force exists because you were pushing against the wall. If you push only a little amount against the wall, the wall will push only a little against you. If you push against the wall with a lot of force, the wall will push against you a lot. Newton also noticed this phenomenon and described it by saying that for every action, there is an equal and opposite reaction. It is now known as the third law of motion.

Newton's third law of motion states that when an object applies a force on another object, the second object applies a force of the same strength on the first object, but the force is in the opposite direction. This idea explains how helicopters and drones work. When a drone pushes the air down, the air pushes the drone up. Air particles may be small and not very massive, but pushing a large amount of air can lead to a large push back.

THREE-DIMENSIONAL THINKING

What **patterns** exist between all forces that you apply to objects or **systems** of objects?

_____
_____
_____
_____
_____
_____
_____
_____

COLLECT EVIDENCE

How do the forces that are present when you push on an object explain what happens when the airboat pushes on the air? Record your evidence (A) in the chart at the beginning of the lesson.

FOLDABLES
Go to the Foldables® library to make a Foldable® that will help you take notes while reading this lesson.

How can you model Newton's third law?

> 📄 **Want more information?**
> Go online to read more about Newton's third law of motion.

All forces come in pairs. Because Newton's third law can be difficult to understand, scientists use models to understand and explain these complicated systems. The free-body diagram model simplifies the many forces involved in a net force.

INVESTIGATION

Back to Back

When air pushes on the sail of a sailboat, the air pushes the boat forward. To represent this, a free-body diagram is made showing a force to the right. Because the object is the boat, not the air, only the force of the air is shown for the object.

A student has come up with an idea for a new airboat. The boat will have the fan attached to the boat facing a sail. The student claims the sail will be full of air, and the boat will sail away very quickly.

1. Draw a free-body diagram of the entire boat system, then draw a diagram with the net force.

EXPLORE/EXPLAIN Lesson 3 Force Pairs

2. Will the boat work as claimed? Explain, using the diagram to support your answer.

___

3. What could be done to improve the student's boat design?

___

Force Pairs The forces described by Newton's third law depend on each other. A **force pair** is the forces two objects apply to each other. Recall that you can add forces to calculate the net force. If the forces of a force pair always act in opposite directions and are always the same strength, why don't they cancel each other? Force pairs are not the same as balanced forces. Balanced forces act on the same object. The force from gravity and the force from the floor act on the same object—you—and are balanced. In force pairs, each force acts on a different object. Look at the ball and the tennis racket below. The ball has the force of the racket pushing it. The racket has the force of the ball pushing on it. The forces do not result in a net force of zero because they act on different objects. Adding forces can only result in a net force of zero if the forces act on the same object.

Normal Force The force that pushes back is sometimes called the normal force. The **normal force** is the force that pushes perpendicular to the object's surface. When you push on the wall, the wall has a normal force that is pushing straight out from the wall. When a tennis ball hits a tennis racket, the racket applies a normal force perpendicular to the racket. This is why a tennis player will turn her racket when she wants to turn the ball to the left or right. The racket applies a normal force in the direction that the player wants the ball to go.

THREE-DIMENSIONAL THINKING

Two students are playing tug of war. Sketch a **model** of three different points along the rope. Draw a model of each end of the rope and of the middle of the rope. Then draw the net force for each point.

Use Newton's third law to construct an **argument** that the net force is the same everywhere along the entire rope.

COLLECT EVIDENCE

How does modeling Newton's third law help explain what happens when an airboat pushes on the air? Record your evidence (B) in the chart at the beginning of the lesson.

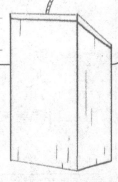

EXPLORE/EXPLAIN Lesson 3 Force Pairs **67**

What happens during a collision?

You might have noticed that if a moving ball hits another ball that is not moving, the motion of each ball changes. When a ball hits other balls, the ball's velocity decreases, and the other balls start moving. How does Newton's third law describe the movement of these balls during a collision?

LAB Bounce Back

Safety

Materials

masking tape   tennis balls (3)   metersticks (2)

Procedure

1. Read and complete a lab safety form.

2. Make a track by using masking tape to secure two metersticks side by side on a table, about 4 cm apart.

3. Place two tennis balls on the track.

   A. Roll one ball against the other. Record your observations below.

   B. Roll the balls at about the same speed toward each other. Record your observations below.

   C. Place the balls so that they touch. Gently roll another ball against them. Record your observations below.

4. Follow your teacher's instructions for proper cleanup.

Data and Observations

A	B	C

68 EXPLORE/EXPLAIN Module: Forces and Motion

Analyze and Conclude

5. Sketch a free-body diagram of each collision in the lab.

A	B	C

6. Using Newton's third law and acceleration, explain why the motion of each of the balls changed.

_____

_____

_____

7. What evidence supports that the forces in both directions were equal? Explain.

_____

_____

_____

8. Imagine that two balls are covered in tape with the sticky side out. When one ball is rolled into the other, they stick together. Model this collision below. How would the motion of the balls change?

EXPLORE/EXPLAIN Lesson 3 Force Pairs

Collision Forces When one object collides with another object, a force is applied to the second object. The second object accelerates in the direction of the force. However, because of Newton's third law, a force is also applied to the first object. As you saw in the Lab *Bounce Back,* when the balls collided, they accelerated in the opposite direction or slowed down. If the mass of the balls was exactly the same, one may have even stopped moving.

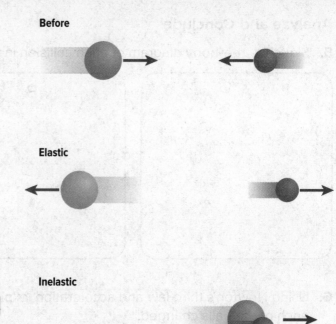

Types of Collisions Objects collide with each other in different ways. When colliding objects bounce off each other, it is an **elastic collision.** If objects collide and stick together, such as when one football player tackles another, the collision is inelastic.

THREE-DIMENSIONAL THINKING
A force is measured by an object's mass and acceleration. If you slowed down a collision between two objects, what effect would that have on the force resulting from Newton's third law?

_____
_____
_____
_____
_____
_____
_____

COLLECT EVIDENCE
What evidence have you discovered to explain the forces during a collision? Record your evidence (C) in the chart at the beginning of the lesson.

A Closer Look: SAFER Barriers

This SAFER barrier has suffered damage from a racecar.

Imagine speeding down the track exceeding speeds of 290 km/h on a cool March day at the Auto Club Speedway in Fontana, California. As you carefully make the third turn, a car speeds in front of you and cuts you off. The next thing you know, you are sliding down the pavement and heading straight into the barrier wall on your car roof! Due to the safety features in the car and the barrier wall, you will walk away from this crash.

Racing scientists and safety crews know that surviving a crash is all about Newton's laws and how to transfer the energy away from the car. The Steel and Foam Energy Reduction (SAFER) barrier is new technology found on oval automobile race tracks and is designed to make racing accidents safer. In 2016, the Auto Club Speedway added the SAFER barriers to its track.

To understand how SAFER walls work, let's look at Newton's third law of motion. As a car hits the barrier wall, it will apply a force to the wall and the wall will apply a force back onto the car. SAFER barriers are designed to lengthen the time it takes to slow the car down, doubling the amount of time a car spends in contact with the track wall. This transfers energy from the car to the track wall instead of to the driver. This transfer can make the difference between a driver being sore and a driver being injured.

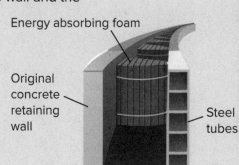

It's Your Turn

Research Most of us will never drive a modern racecar. Most of us do ride in cars though. Use the Internet or other sources to research one of the safety features developed by a progress in science that is built into modern roads to increase safety for drivers and passengers. Prepare a multimedia presentation to share your findings with the class. Be sure to include how Newton's third law is used in the safety feature.

LESSON 3 Review

Summarize It!

1. **Model** Make a model that shows how Newton's third law describes the forces acting between any two objects in an unbalanced force pair.

Three-Dimensional Thinking

A person is applying a force to the right on an object as shown. Use the model below to answer questions 2 and 3.

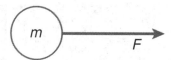

2. What forces are acting on the person?

    A a slightly smaller force to the left because the object is accelerating

    B a force equal to the force applied going to the left

    C a force to the right to apply the force to the object

    D a force to the right because the object is accelerating

3. The person is standing on ice with little to no friction. What will be the motion of the person applying the force to the object?

    A begin to move to the right because that is the direction of the push

    B no change in motion because the person is pushing the object

    C begin to move to the left because the object pushes on the person

    D begin to move to the right with the object

4. Which of the following systems does NOT represent a force pair?

    A When you push on a bike's brakes, the friction between the tires and the road increases.

    B When a diver jumps off a diving board, the board pushes the diver up.

    C When an ice skater pushes off a wall, the wall pushes the skater off of the wall.

    D When a boy pulls a wagon, the wagon pulls back on the boy.

Real-World Connection

5. **Explain** When you run, your feet are pushing you forward. Friction keeps your foot in contact with the ground. According to Newton's third law, you are pushing the ground back. Construct an explanation for why Earth is not changing its motion.

_____

_____

_____

_____

6. **Explain** To steer an airboat, rudders sit behind the fan. When the air passes through the turned rudders, it turns the boat. Use Newton's third law to construct an explanation on how the rudders turn the boat.

_____

_____

_____

 Still have questions?
Go online to check your understanding about force pairs and Newton's third law

REVISIT SCIENCE PROBES Do you still agree with the statement you chose at the beginning of the lesson? Return to the Science Probe at the beginning of the lesson. Explain why you agree or disagree with that statement now.

EXPLAIN THE PHENOMENON

Revisit your claim about how pushing the air moved the airboat. Review the evidence you collected. Explain how your evidence supports your claim.

KEEP PLANNING
STEM Module Project Engineering Challenge

Now that you've learned about Newton's third law, go to your Module Project to sketch your technology, the components in the system, and the forces. Keep in mind how force pairs will be a constraint.

LESSON 4 LAUNCH

Ball Toss

Jose tosses a ball high into the air. The ball eventually comes down so his friends can catch it. Jose and his friends have different ideas about why the ball comes back down. This is what they said:

Jose: I think it comes down because Earth is pulling on it.
Eddie: I think it comes down because it runs out of force.
Lucy: I think it comes down because air pressure pushes it down.
Dinah: I think it comes back down because no forces are acting on it.

Who do you most agree with? _____ Explain your thinking about why the ball comes back down.

_____
_____
_____
_____
_____
_____

You will revisit your response to the Science Probe at the end of the lesson.

LESSON 4
Gravitational Force

ENCOUNTER THE PHENOMENON | What pulls the skydiver to the ground?

Hold your pencil out in front of you and let go. What happens to the pencil? Record your observations in the space below.

Now hold both a book and a pencil out in front of you. Let go of the book and the pencil at the same time. What happens? Record your observations in the space below.

GO ONLINE
Watch the video *Skydiving* to see this phenomenon in action.

ENGAGE Lesson 4 Gravitational Force

EXPLAIN
THE PHENOMENON

You already know that a force must be applied to cause a change in motion. Just like the objects you used in the activity, when a skydiver leaves the plane, she moves downward. Make a claim about what you think caused the skydiver to fall to the ground.

CLAIM
The skydiver falls down because...

 COLLECT EVIDENCE as you work through the lesson. Then return to these pages to record your evidence.

EVIDENCE
A. What evidence have you discovered to explain why the skydiver fell downward?

B. What evidence have you discovered to explain the relationship between mass and gravity and the relationship between distance and gravity?

78 Module: Forces and Motion

MORE EVIDENCE

C. What evidence have you discovered to explain why the skydiver did not move toward the plane?

When you are finished with the lesson, review your evidence. If necessary, based on the evidence, revise your claim.

REVISED CLAIM

The skydiver fell because...

Finally, explain your reasoning for how and why your evidence supports your claim.

REASONING

The evidence I collected supports my claim because...

Lesson 4 Gravitational Force

How can you change an object's motion without touching it?

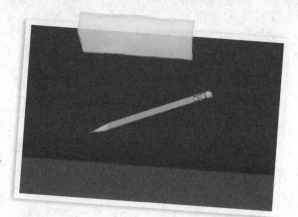

In the previous lessons, an object applying a force had to be touching another object to cause a change in motion. But what caused the pencil to change motion? What other forces act like this?

Want more information?
Go online to read more about gravitational forces.

FOLDABLES
Go to the Foldables® library to make a Foldable® that will help you take notes while reading this lesson.

LAB Use the Forces

Safety

Materials

magnets (2)   balloon   wool cloth
packing peanuts (20)   paper clips (20)   paper

Procedure

1. Read and complete a lab safety form.

2. Examine the materials. Investigate to see if you can change the motion of the paper clips or the packing peanuts without another object physically touching them. Record your observations in your Science Notebook.

3. Follow your teacher's instructions for proper cleanup.

Analyze and Conclude

4. Record what you did to change the motion of the paper clips or the packing peanuts.

---

80 EXPLORE/EXPLAIN Module: Forces and Motion

5. How are these forces similar to or different from what caused the pencil from the beginning of the lesson to fall?

_____

_____

_____

Noncontact Forces A force must have been acting on the pencil to pull it down. The source of this force did not need to physically touch an object to cause it to fall. This force is different from the forces you have learned about previously because it is a noncontact force. A **noncontact force** is a force that one object can apply to another object without touching it. Electric forces can cause someone's hair to change direction when a charged balloon is brought near the hair as shown in the figure on the right. The balloon does not have to touch the hair to cause it to move. Magnets can move paper clips even though the magnet is not in contact with the paper clip.

What pulls things down?

In the Lab *Use the Forces,* you may have moved the paper clips with the magnet. Or using the wool cloth and the balloon, you may have created an electric force that caused the packing peanuts to move. The source of a force must be an object. What object created the force that caused the pencil to fall down?

INVESTIGATION

The Pencil Dropped Around the World

A classroom of students noticed that a pencil fell downward in their classroom in California. To find out if a pencil always falls downward at all locations on Earth, they collected data from other classrooms around the world. Each participating classroom was asked to drop a pencil from a height of one meter.

EXPLORE/EXPLAIN Lesson 4 Gravitational Force **81**

They recorded the direction the pencil moved and the time it took for the pencil to hit the ground. The following table is the data collected from each classroom.

Location	Direction of pencil movement	Time (s)
California, USA	downward	0.450
South Africa	downward	0.454
Russia	downward	0.453
China	downward	0.452
Brazil	downward	0.451
Sweden	downward	0.450

1. Using evidence from the table, what logical conclusion can you draw about the source of the force that acted on the pencils?

2. Was the size of the force acting on each pencil the same or different? Explain.

Gravitational Force Because all of the pencils fell downward no matter the location on Earth, Earth must be the source of the force. The magnitude of the force was also the same on all the pencils, so this force must act equally on similar objects. This force is called gravitational force or gravity. **Gravity** is an attractive force that exists between all objects that have mass. Mass is the amount of matter in an object. Both the pencil and Earth have mass, so both the pencil and Earth exert a gravitational pull on each other. Because an object does not have to be touching Earth for gravity to act on it, the force of gravity must be exerted through space as a field. A **field** is a region of space that has a physical quantity (such as a force) at every point. The gravitational field of Earth surrounds Earth at all points. A small gravitational field also surrounds the pencil.

COLLECT EVIDENCE

How does the existence of gravity explain why the skydiver fell downward? Record your evidence (A) in the chart at the beginning of the lesson.

What factors affect the strength of a gravitational force?

A pencil has less mass than Earth. Do they exert the same amount of force on each other? What if there was a pencil on the *International Space Station?* Would Earth exert the same amount of force on it as the pencil on Earth's surface?

INVESTIGATION

The Force of Gravity

How could you increase the force of gravity?

GO ONLINE Explore the PhET interactive simulation *Gravity Force Lab*.

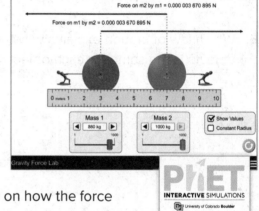

After exploring the simulation on your own, reset the simulation and follow the instructions below.

1. Determine how to change the force of gravity on the object on the right. In the graphic organizer below, circle the words that best describe your observations on how the force of gravity on the object on the left changes.

> When the force of gravity changes on one object, the force of gravity on the second object always changes by
>
>   the same amount.   a different amount.

2. Move the object on the right closer to and farther away from the object on the left.

3. Complete the graphic organizer below using *increases* or *decreases* to explain how changing distance affects the force of gravity on the two objects.

| Decreasing the distance between the two objects | _____ the force of gravity on the first object.
 _____ the force of gravity on the second object. |

| Increasing the distance between the two objects | _____ the force of gravity on the first object.
 _____ the force of gravity on the second object. |

EXPLORE/EXPLAIN Lesson 4 Gravitational Force

4. Next, increase the mass of the object on the right. Then decrease the mass of the object on the right.

5. Complete the graphic organizer using *increases* or *decreases* to explain how changing mass affects the force of gravity on the two objects.

Decreasing the mass of one of the two objects	_____ the force of gravity on the first object. _____ the force of gravity on the second object.
Increasing the mass of one of the two objects	_____ the force of gravity on the first object. _____ the force of gravity on the second object.

6. Which law of motion explains why the force of gravity between two objects is the same for each object regardless of its mass? Explain.

_____

_____

_____

Gravitational Force and Mass When the mass of one or both objects increases, the gravitational force between them also increases. This is a direct proportional relationship. Each object exerts the same attraction on the other object.

If both objects exert the same attraction on each other, why does it look like the pencil falls toward Earth? The effect of mass on the force of gravity is most noticeable when one object is very massive, such as a planet, and the other object has much less mass, such as a pencil. Even though the force of gravity acts equally on both objects, the less massive object accelerates more quickly due to its smaller mass. Because the planet accelerates so slowly, all you observe is the object with less mass "falling" toward the object with greater mass.

Gravitational Force and Distance As the distance between objects increases, the attraction between objects decreases. This is an inverse proportional relationship as seen in the graph on the right. For example, if your mass is 45 kg, the gravitational force between you and Earth is about 440 N. On the Moon, about 384,000 km away, the gravitational force between you and Earth would only be about 0.12 N.

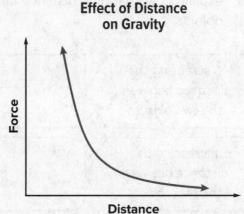

Effect of Distance on Gravity

84 EXPLORE/EXPLAIN Module: Forces and Motion

Read a Scientific Text

HISTORY Connection In the late 1600s, Sir Isaac Newton, an English scientist and mathematician, studied the effects of gravity on objects on Earth and in space. From these studies he wrote *Newton's Principia. The Mathematical Principles of Natural Philosophy*. In the text he summarized the three laws that govern all motion and the law of universal gravitation.

CLOSE READING

Inspect
Read the passage *Newton's Principia. The Mathematical Principles of Natural Philosophy*.

Find Evidence
Reread the passage. Underline the evidence Newton presented for his concept of gravity.

Make Connections
Communicate With your partner, discuss if this evidence would be enough to convince you that gravity exists between all objects.

PRIMARY SOURCE

Passage from: Newton's Principia. The Mathematical Principles of Natural Philosophy by Sir Isaac Newton

Lastly, if it universally appears, by experiments and astronomical observations, that all bodies about the earth gravitate towards the earth, and that in proportion to the quantity of matter which they severally contain, that the moon likewise, according to the quantity of its matter, gravitates towards the earth; that, on the other hand, our sea gravitates towards the moon; and all the planets mutually one towards another; and the comets in like manner towards the sun; we must, in consequence of this rule, universally allow that all bodies whatsoever are endowed with a principle of mutual gravitation. For the argument from the appearances concludes with more force for the universal gravitation of all bodies than for their impenetrability; of which, among those in the celestial regions, we have no experiments, nor any manner of observation. Not that I affirm gravity to be essential to bodies: by their *vis insita* [power implanted or inertia] I mean nothing but their *vis inertiae* [weak force]. This is immutable. Their gravity is diminished as they recede from the earth.

Source: Newton's Principia. The Mathematical Principles of Natural Philosophy by Sir Isaac Newton

THREE-DIMENSIONAL THINKING

On the figure below, add arrows to **model** the size and direction of the force of gravity for the missing forces in each **system** below.

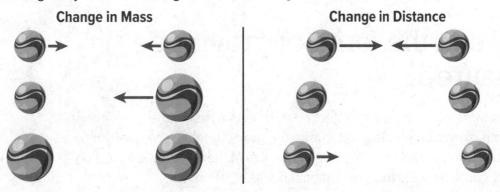

EXPLORE/EXPLAIN Lesson 4 Gravitational Force **85**

EARTH SCIENCE Connection Gravity does not just cause pencils to fall to Earth. It is involved in the cycling of water on Earth, and it governs the movements and patterns of the entire visible universe from the Moon orbiting Earth, to asteroids and comets orbiting the Sun, and to entire galaxies.

The force of gravity played a major role in the formation of the solar system. The solar system formed from a cloud of gas, ice, and dust. Gravity pulled the materials closer together. The cloud shrank and flattened into a rotating disk. The center of the disk became denser, forming a star—the Sun. The planets began to take shape from the remaining bits of material. Earth formed as gravity pulled these small particles together. As they collided, they stuck to each other and formed larger masses. The larger objects had more mass, so their gravity grew. This caused them to attract more particles. Eventually enough matter collected and formed Earth.

Earth travels around the Sun due to the gravitational force between the Sun and Earth. Everything in our solar system is held in orbit due to the gravitational force created by the Sun. If the gravitational force disappeared, Earth would break away from its orbit and continue traveling in a straight line into space.

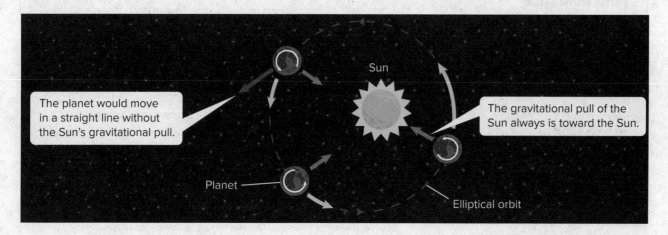

COLLECT EVIDENCE
How does the relationship between mass and gravity and the relationship between distance and gravity explain why the skydiver fell as she did? Record your evidence (B) in the chart at the beginning of the lesson.

How can the force of gravity be measured?

Earth has more mass than any object near you. As a result, the gravitational force Earth exerts on you is greater than the gravitational force exerted by any other object. How can you measure the gravitational force exerted on an object caused by Earth's gravitational pull? Let's find out!

LAB Weighing Washers

Safety

Materials
metal washers (10)
balance
spring scale

Procedure

1. Read and complete a lab safety form.

2. Use a balance to find the mass of 4 metal washers. Record the mass in the Data and Observations section below. Note the units.

3. Hang the washers from the hook on a spring scale. Record the force. Note the units.

4. Repeat steps 2 and 3 with 6, 8, and 10 washers.

5. Follow your teacher's instructions for proper cleanup.

Data and Observations

Number of washers	Mass (units = _____)	Force (units = _____)
4		
6		
8		
10		

Analyze and Conclude

6. How did the spring scale measure the force of gravity on the washers?

_____
_____
_____
_____

EXPLORE/EXPLAIN Lesson 4 Gravitational Force

Analyze and Conclude, continued

7. **MATH Connection** Divide each force of gravity by its mass to see if there is a common ratio between mass and weight. Show your work in the space below.

8. **MATH Connection** Did you see a pattern in the ratios between mass and weight? Explain what you think this pattern means.

_____

_____

_____

Gravitational Acceleration Recall that Newton's second law of motion states that the force on an object can be found by multiplying mass by acceleration. When you divide the force of gravity on the washers by the mass of the washers, you can find the gravitational acceleration.

Did you notice the ratio between each force of gravity and the mass of the washers was similar and did not change with mass? When the only force acting on a falling object is gravity, all objects fall with the same acceleration. Close to Earth's surface, the acceleration of these objects is 9.8 m/s². This is gravitational acceleration (*g*). Gravitational acceleration is why, even though the pencil and the book from the beginning of the lesson were different masses, they hit the ground at the same time. Even if the mass of an object increases, the force of gravity between the object and Earth will also increase by the same amount. This is why the acceleration is always the same if no additional forces are acting on the object. The time-lapse photograph on the right shows a feather and a ball with two different masses dropped at the same time. This will only happen in a vacuum when there is no air resistance. Air resistance opposes gravity as an object falls.

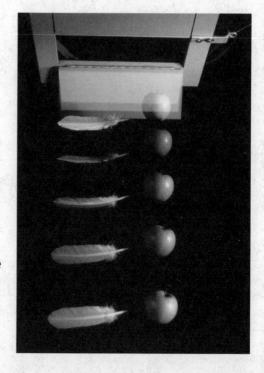

Weight Even though the washers were not falling, the force due to Earth's gravity still was pulling them downward. **Weight** is the gravitational force exerted on an object. Near Earth's surface, an object's weight is the gravitational force exerted on the object by Earth. Because weight is a force, it is measured in newtons (N). Near Earth's surface, the weight of an object in newtons is about ten times its mass in kilograms. The weight of an object can be calculated by using the following equation.

Weight Equation

weight (N) = mass (kg) × gravitational acceleration (m/s²)

$$W = mg$$

Normal Force A pencil sitting on a table is acted on by gravity. This force in the downward direction is equal to its weight. Because gravity is acting on the pencil, the pencil is pushing down on the table. Recall, Newton's third law of motion states that for every force there is an opposite equal force. The table pushes up on the pencil with an equal but opposite force. The force of the table on the pencil is referred to as the normal force.

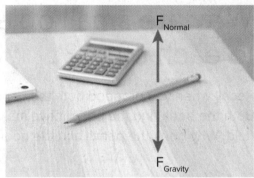

MATH Connection For each of the systems below, calculate the weight of the object. Then, apply scientific understanding to draw a free-body diagram of the system.

A. A book with a mass of 2 kg is at rest on top of a desk.

B. A pencil with a mass of 0.0075 kg is falling toward the floor.

THREE-DIMENSIONAL THINKING

Construct an argument on whether the weight of a pencil would change as the pencil falls from 10 m to the ground.

_____
_____
_____
_____

Why are small objects not attracted to each other?

Recall when you dropped both the pencil and the book at the same time. Both the book and the pencil have mass, so each has its own gravitational field. Why don't the pencil and the book attract each other?

INVESTIGATION

Gravity of Objects

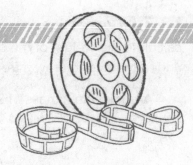

▶ **GO ONLINE** Watch the animation *Gravity of Objects* to explore the gravitational force made by the book and the pencil.

Why doesn't a book cause a pencil to move toward it when the pencil is placed close by?

_____
_____
_____

Gravity of Small Objects All objects have mass, so they all have a gravitational field. The masses of the pencil and the book are very small compared to Earth's, so the gravitational field of each is very small. They pull on each other slightly, but not enough to cause a change in motion.

COLLECT EVIDENCE

How does the gravitational attraction between all objects explain why the skydiver does not move toward the plane? Record your evidence (C) in the chart at the beginning of the lesson.

EXPLORE/EXPLAIN Module: Forces and Motion

A Closer Look: Space Travel

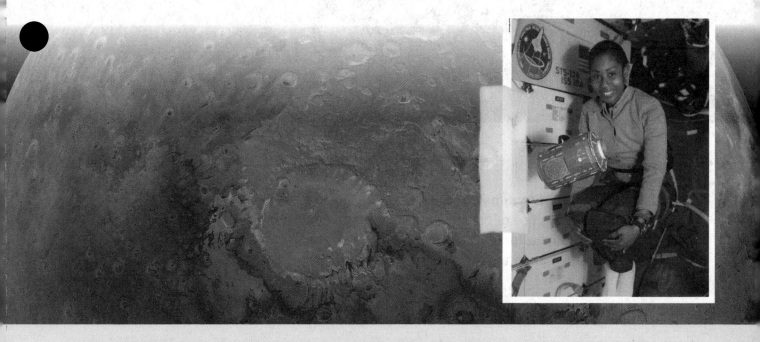

Gravity is all around us every day. It is constantly pulling on us as we walk the surface of Earth. On a mission to Mars, this constant gravity would be gone. How does the loss of Earth's gravity affect our bodies?

- Escaping Earth's gravitational field can cause major motion sickness. As you move from one gravity field to another, your spatial orientation needs to readjust. This affects your balance, locomotion, and hand-eye coordination.

- Without Earth's gravity acting on your body, your bones lose density at the rate of 1% a month. On Earth the rate is only 1–1.5% a year. Because it does not take a lot of effort to float in space, without exercising you could lose muscle strength and endurance.

- In space there is nothing to pull the fluids in your body downward. This could cause fluids to pool at the top of your body. This could cause an increased pressure on your eyes which would affect your vision.

National Aeronautics and Space Administration (NASA) is studying these effects and more on human bodies as they develop their mission to send astronauts to Mars.

It's Your Turn

WRITING Connection How are the uses of technologies limited by the findings of scientific research? Research the solutions NASA is currently developing to combat the effects of the loss of gravity on the human body. Choose one solution and create a blog on how it will help astronauts on a mission to Mars.

LESSON 4
Review

Summarize It!

1. **Construct and Present Arguments** You have been asked to join a debate about the existence of gravity. Develop an argument to support the idea that gravitational forces are attractive and depend on the mass of the object. Use evidence to add validity to your argument.

2. **Design a Solution** You have been asked to design a solution that would make moving heavy furniture easier. Plan an investigation to show how well your solution works.

Three-Dimensional Thinking

The model below represents a star orbited by two planets—Planet A and Planet B. The star is also orbited by a mysterious object, Object X, which entered into the star's gravitational field. The star is the most massive object, followed by Planet B, Planet A, and Object X. Use the model to answer questions 3–4.

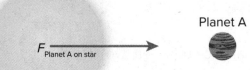

3. In the model above, how is the mass of the objects represented?

    A The mass is represented by the size of the objects.

    B The mass is represented by the distance between the objects.

    C The mass is represented by the color of the objects.

    D The mass is not represented.

4. The gravitational force from Planet A on the star is shown in the model. How should the arrow that represents the gravitational force from the star on Planet A be represented?

    A It should point from Planet A toward the star and will be longer because the star has more mass.

    B It should point from Planet A toward the star and will be the same size because it is an equal and opposite force.

    C It should point from Planet A toward the star and will be shorter because Planet A has less mass.

    D There is no arrow to represent because gravitational force is only in one direction.

EVALUATE Lesson 4 Gravitational Force

Real-World Connection

5. **Predict** If an astronaut moved away from Earth in the direction of the Moon, how would the gravitational force between Earth and the astronaut change? How would the gravitational force between the Moon and the astronaut change?

_____
_____
_____
_____

6. **Explain** You overhear someone say the gravitational force between two 50-kg objects is less than the gravitational force between a 50-kg object and a 5-kg object. What question could you ask this person in order to challenge their argument? Explain.

_____
_____
_____
_____

 Still have questions?
Go online to check your understanding about gravitational forces.

REVISIT PAGE KEELEY SCIENCE PROBES

Do you still agree with the student you chose at the beginning of the lesson? Return to the Science Probe at the beginning of the lesson. Explain why you agree or disagree with that student now.

PLAN AND DESIGN

STEM Module Project Engineering Challenge

Now that you've learned about gravity, go to your Module Project to model your new technology. Keep in mind the forces acting on each object before and after the collision.

EXPLAIN THE PHENOMENON

Revisit your claim on why the pencil and the book both fell down at the same time. Review the evidence you collected. Explain how your evidence supports your claim.

STEM Module Project
Engineering Challenge

Crash Course

Like the rowers at the beginning of the module, a car will continue to move unless acted on by a force. If the brakes fail, the car will remain in motion. The mayor of a nearby town has hired your group of engineers to develop a solution to help protect drivers and reduce the damage that might result from a car or other vehicle striking a new bridge support. The mayor would like your group to design, test, and evaluate a possible solution to the problem. In addition, the mayor needs you to provide results of investigations to show that the change in motion of a vehicle depends on the sum of forces and mass to help inform other city leaders of the science behind the solution.

Planning After Lesson 1

Define the problem. Include in the group's solution:

- who needs this solution
- what needs must be met
- any relevant scientific issues
- any societal or environmental impacts

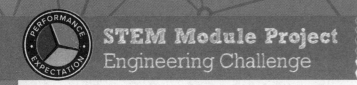

Planning After Lesson 1, continued

Explain the importance of why all the components on the previous page must be defined.

What criteria does this technology need to meet?

The technology your group has been requested to design clearly has safety requirements. What factors about the object and its motion must be considered? What other constraints are you limited by?

Planning After Lesson 2

Consider the Challenge with respect to Newton's second law of motion. Describe the desired goal for the motion of the vehicle, including its acceleration.

What factors affect acceleration, and how will they impact your decisions in your engineering challenge?

How do you predict balanced forces and unbalanced forces will be involved in your design?

Planning After Lesson 3

Research technologies that are used to reduce the damage of vehicles hitting road structures. Record the questions and answers that drive your research in the space below. Include citations for your sources.

STEM Module Project
Engineering Challenge

Planning After Lesson 3, continued

Create a table that lists:

- each type of technology that was researched
- how the technology reduces damage from crashes
- the societal and environmental impacts of each technology

Add columns to the table that identify each criteria and constraint. Use the table to evaluate the strengths and weaknesses of each technology.

Begin designing the technology. Sketch the technology. In the sketch include the components in the system, the forces involved, the technologies and materials that will be used, how Newton's third law has been applied to the solution, and how this technology provides the best solution to the problem by addressing the criteria and constraints of the problem. Once a sketch has been approved by your teacher, start building the technology as a group.

STEM Module Project
Engineering Challenge

Planning After Lesson 4

Model the system of objects in your solution as a set of masses, identifying the mass of each relevant object and the relative magnitude and direction of the forces acting on them, including gravitational forces.

Create Your Design

As a group, develop a plan on how to investigate which technology designed by you and your classmates best meets the criteria and constraints. In this plan include:

- What criteria and constraints will be used in the evaluation of the design solution?
- What data are needed to provide evidence for the object being subjected to balanced forces and unbalanced forces?
- How will you determine and measure the factors of the motion of the object (including its reference frame and the appropriate units for distance and time), the mass of the object (including units), and the forces acting on it (including balanced and unbalanced forces)?
- How will you use your knowledge of Newton's third law to determine how well the design solution meets the criteria and constraints?

Create tables to organize the data before testing the devices developed by your class. Once your tables are made, begin testing your technologies!

Analyze and Conclude

Analyze the data collected by your class. Compare the solutions based on the results of their performance against the defined criteria and constraints. Explain which characteristics from the class's technologies best met the criteria and constraints of the project. Make a claim about the relative effectiveness of the proposed solution based on the strengths and weaknesses of each. Support your claim with evidence, and present your argument to the class.

Identify the value of the technology you designed for society.

Describe how the choice of technologies used in the design is affected by the constraints of the problem and the limits of technological advances.

Congratulations! You've completed the Engineering Challenge requirements!

Module Wrap-Up

REVISIT THE PHENOMENON

Think about everything you have learned in the module about how forces determine the motion of an object. Construct an explanation about why the boat can move without anyone rowing.

OPEN INQUIRY

What are one or two questions you still have about the phenomenon?

_____

_____

_____

_____

Choose the question that interests you the most. Plan and conduct an investigation to answer this question.

EVALUATE Module: Forces and Motion

Mechanical Energy

ENCOUNTER
THE PHENOMENON

What role does energy play in an amusement park ride?

Put your hands in the air!

GO ONLINE
Watch the video *Put your hands in the air!* to see this phenomenon in action.

Communicate Think about what you know about energy. With a partner, discuss your thoughts. Then record what you both would like to share with the class below.

Module: Mechanical Energy

STEM Module Project Launch
Science Challenge

Energy at the Amusement Park

The CEO of the local amusement park has heard that energy is lost during a run of the park's vertical-drop ride. He wonders if they could collect the energy to power the park's generator. Your company has been contracted to develop a model detailing the change in position, speed, forces, and types of energy during a run of the ride. With your team, research different vertical-drop rides. Next, develop your model. Finally, construct an argument on whether or not energy can be transferred to power a generator for the park.

Lesson 1 Kinetic Energy

Lesson 2 Potential Energy

Lesson 3 Conservation of Energy

Start Thinking About It

How do you think energy is associated with the vertical-drop ride shown in the photo above? Discuss your thoughts with your group.

STEM Module Project
Planning and Completing the Science Challenge
How will you meet this goal? The concepts you will learn throughout this module will help you plan and complete the Science Challenge. Just follow the prompts at the end of each lesson!

LESSON 1 LAUNCH

Soccer Ball

Five soccer players argued about when a soccer ball has energy. This is what they said:

Jorge: The ball has to be moving to have energy.
Kurt: The ball has energy only at the moment it is kicked.
Amos: The ball has energy only when it is not moving.
Alan: The ball has energy when it is moving and not moving.
Flavio: The ball has no energy. There is no source of energy in the ball.

Who do you agree with the most? _____
Explain your thinking. What rule or reasoning did you use to decide when the soccer ball has energy?

_____
_____
_____
_____
_____
_____

You will revisit your response to the Science Probe at the end of the lesson.

LESSON 1
Kinetic Energy

ENCOUNTER THE PHENOMENON

What determines how far you can move a ball?

Watch your teacher demonstrate how far a paper cup can move. Next, create your own setup and try to make the paper cup move farther. Record your setup and the distance the cup moved in the space below.

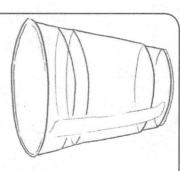

GO ONLINE
Watch the video *Going, Going, Gone* to see this phenomenon in action.

ENGAGE Lesson 1 Kinetic Energy

EXPLAIN THE PHENOMENON

The baseball connects with the bat. Will the ball stay in the infield? Will it soar over the crowd and out of the stadium? Either way, the factors that make the ball move are the same. Make a claim about what determines how far an object, such as the ball, will travel.

CLAIM
The distance an object, such as a ball, travels is determined by...

 COLLECT EVIDENCE as you work through the lesson. Then return to these page to record your evidence.

EVIDENCE
A. What evidence have you discovered to explain how kinetic energy and mass are related?

Module: Mechanical Energy

MORE EVIDENCE

B. What evidence have you discovered to explain how kinetic energy and speed are related?

When you are finished with the lesson, review your evidence. If necessary, based on the evidence, revise your claim.

REVISED CLAIM

The distance an object, such as a ball, travels is determined by...

Finally, explain your reasoning for how and why your evidence supports your claim.

REASONING

The evidence I collected supports my claim because...

What causes motion?

Think back to the activity at the beginning of the lesson. When the sphere hit the back of the cup, the cup moved. That was a change in position caused by the motion of the sphere. What did you do to increase the distance the cup moved? Let's investigate more to find out.

Want more information?
Go online to read more about kinetic energy.

FOLDABLES
Go to the Foldables® library to make a Foldable® that will help you take notes while reading this lesson.

INVESTIGATION

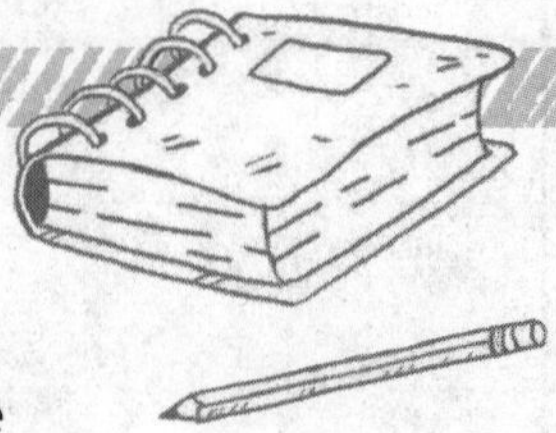

Rolling On

1. Record three setups that other classmates used to increase the distance the cup moved.

Setup A	Setup B	Setup C

2. Repeat the task with each setup. Record your observations below.

3. Evaluate which setup caused the cup to move the farthest and why.

_____
_____
_____

112 EXPLORE/EXPLAIN Module: Mechanical Energy

Kinetic Energy The change in motion of the cup occurred because the sphere had a form of energy. This form is called kinetic energy. **Kinetic energy** (KE) is the energy due to motion. When the baseball flies across the field, it also has kinetic energy associated with its motion. All moving objects have kinetic energy. You may have changed how much speed the sphere had before it hit the cup or you may have used a sphere with a different mass. You will investigate how these factors affect the kinetic energy of an object.

How are kinetic energy and mass related?

Have you ever thrown a baseball or a tennis ball? A baseball has a lot more mass than a tennis ball. How does this affect the amount of kinetic energy the ball has? Let's investigate!

Safety

Materials

spheres (3)   meterstick
balance   dropper
clay   water

Procedure

1. Read and complete a lab safety form.

2. Measure the mass of each sphere and record the mass in a data table you create in the Data and Observations section on the next page.

3. Hold all three spheres 1 m above a pad of clay. Drop them at the same time into the clay.

4. Measure the volume of each indent, or crater, by filling it with water from a dropper. Record the volume of water in the data table.

5. Repeat steps 3–4 to record two more trials.

6. Follow your teacher's instructions for proper cleanup.

EXPLORE/EXPLAIN Lesson 1 Kinetic Energy 113

Data and Observations

Analyze and Conclude

7. **MATH Connection** Find the average crater volume for each sphere.

_____

_____

8. **MATH Connection** What patterns do you notice in the data? Write a ratio to describe the pattern.

_____

_____

9. Would an object with the most or the least kinetic energy result in the largest crater volume? Explain.

_____

_____

_____

10. Plot the mass and average crater volume on the grid below. Plot mass on the horizontal axis and volume on the vertical axis. Draw a line that goes through the most points. Label the axes and add a title to your plot.

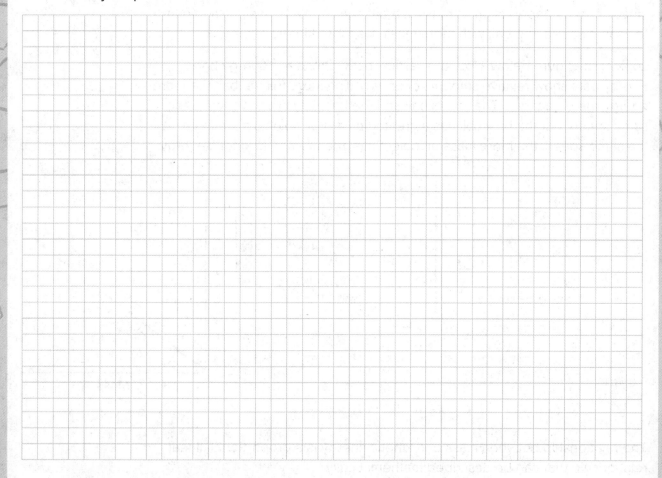

11. **MATH Connection** Analyze the data you collected to interpret the relationship between mass and volume. Write the equation that describes this relationship.

_____
_____
_____
_____

12. Using the data, explain how mass affects kinetic energy.

_____
_____
_____
_____

Kinetic Energy and Mass Did you notice in the Lab *Mass Matters* that the sphere with the largest mass created the largest crater? As it fell it had kinetic energy. When it hit the clay it created a larger crater because it had more kinetic energy than the spheres with less mass. A moving object's kinetic energy depends on its mass. If a baseball and a tennis ball move at the same speed, the ball with more mass has more kinetic energy.

Look at the figure below. Note the vertical bars. These are energy bars. Energy bars show relative amounts of energy. The more full the bar, the more energy the object has. The tennis ball and the baseball are traveling at the same speed, but the baseball has a greater mass. For objects traveling at the same speed, the more mass an object has, the greater its kinetic energy.

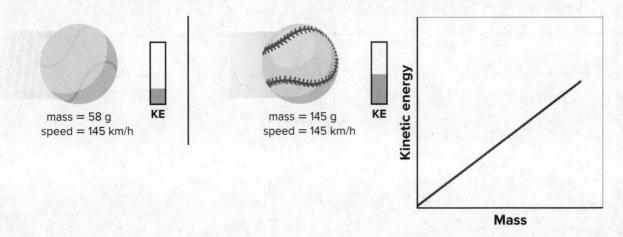

The relationship between mass and kinetic energy is a proportional, linear relationship that can be described mathematically.

$$KE \propto m$$

THREE-DIMENSIONAL THINKING

For each system determine how the quantity of kinetic energy would increase or decrease.

1. If the mass of an object is increased by a factor of 4, kinetic energy increases by a factor of _____.

2. If the mass of an object is decreased by a factor of $\frac{1}{2}$, kinetic energy decreases by a factor of _____.

COLLECT EVIDENCE

How does the mass of the ball determine how far the ball will travel? Record your evidence (A) in the chart at the beginning of the lesson.

EXPLORE/EXPLAIN Module: Mechanical Energy

How are kinetic energy and speed related?

When the average person throws a baseball, it can reach speeds of 64 kilometers per hour. Some pitchers can throw a baseball at speeds over 145 kilometers per hour! Do you think each baseball thrown has the same amount of kinetic energy? Let's investigate!

LAB Picking Up Speed

Safety

Materials

sphere
balance
clay
meterstick
dropper
water

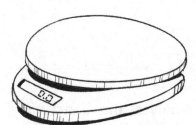

Procedure

1. Read and complete a lab safety form.

2. Measure the mass of the sphere and record the mass in a data table you create in the Data and Observations section on the next page.

3. Hold the sphere 0.2 m above a pad of clay. Drop it into the clay.

4. Measure the volume of the indent, or crater, by filling it with water from a dropper. Record the volume of water in the data table.

5. Repeat step 3 from two additional heights, 0.82 m and 1.84 m.

6. Repeat step 4 for each height.

7. Conduct two more trials for each height.

8. Follow your teacher's instructions for proper cleanup.

Data and Observations

Analyze and Conclude

9. **MATH Connection** Find the average crater volume for each height.

_____

_____

_____

10. What patterns do you notice in the data?

_____

_____

_____

11. How does changing the height from where you dropped the sphere change the motion of the sphere? Explain.

_____

_____

_____

12. Plot the height and average crater volume on the grid below. Plot the height on the vertical axis and the volume on the horizontal axis. Label the axes and add a title to your plot. Draw a line that goes through most of the points.

13. **MATH Connection** Analyze the data you collected to interpret the relationship between height and volume.

_____

_____

14. Compare your data with data from the sphere of the same mass in the Lab *Mass Matters*. Argue which factor, mass or velocity, has a greater influence on kinetic energy. What is your evidence?

_____

_____

_____

_____

EXPLORE/EXPLAIN Lesson 1 Kinetic Energy

Kinetic Energy and Speed In the Lab *Picking Up Speed,* you observed that an object's kinetic energy also depends on its speed. The faster an object moves, the more kinetic energy it has. Look at the figure below. The pitcher used the same softball, but the speed was different for the two pitches. The faster pitch has more kinetic energy. This is represented in the energy bars.

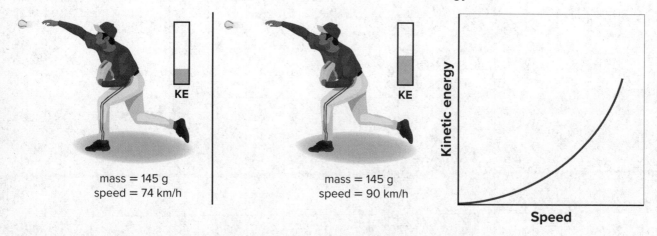

There is a proportional relationship between kinetic energy and speed. Did you notice that the speed-volume graph in the Lab *Picking Up Speed* was different from the mass-volume graph in the Lab *Mass Matters*? This is because even though both are proportional relationships, they are not the same type of proportional relationship.

When you increase the mass of an object, the kinetic energy increases by the same factor. When you increase the speed of an object, the kinetic energy increases by the square of the factor. This is a square, nonlinear relationship, which is shown in the graph above. A ball gains kinetic energy when thrown. If a pitcher throws the ball twice as fast, this quadruples its kinetic energy. This relationship can be expressed mathematically by the following expression:

$$KE \propto v^2$$

 THREE-DIMENSIONAL THINKING
A horse galloped at 40 km/h, then it began to slow down. **Construct an explanation** about what happens to its kinetic energy as the horse slowed to 20 km/h. Explain using evidence.

_____

_____

_____

COLLECT EVIDENCE
How does the speed of the ball determine how far the ball will travel? Record your evidence (B) in the chart at the beginning of the lesson.

A Closer Look: Kinetic Energy

When a ball is thrown or rolled, the ball moves in two ways. So far you have investigated the motion energy due to change of position. This is translational kinetic energy. The ball moved from one point to another.

The second type of motion energy does not involve changing position. Think about a Ferris wheel. It doesn't have translational kinetic energy because the ride does not move from one part of the park to another. However, it is rotating. The energy associated with the rotational motion is rotational kinetic energy.

An object can have both translational and rotational kinetic energy. For instance, a bowling ball rolling down a bowling lane is changing position. The ball leaves the bowler's hand and travels toward the pins. If you watch carefully, you will notice that the finger holes aren't always visible. This is because the ball is rotating. The bowling ball has translational and rotational kinetic energy as it moves down the bowling lane.

It's Your Turn

Discover Besides a ball, identify another object or action that has both translational and rotational energy. Find or make a short video clip of what you've discovered. Include your video clip in a presentation as evidence of the two types of kinetic energy.

LESSON 1
Review

Summarize It!

1. **Organize** Complete the graphic organizer below to describe the relationships of the variables that result in kinetic energy.

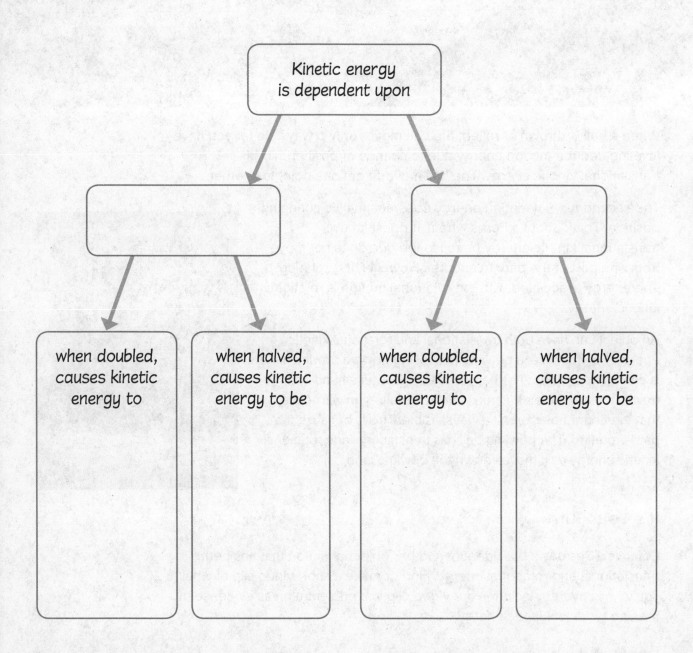

122 EVALUATE Module: Mechanical Energy

Three-Dimensional Thinking

2. Aiden collected canned goods for a neighborhood service project. He pulled a plastic wagon behind him to put the items in. From the time Aiden began until he finished collecting, the mass of the wagon tripled. The walk back to Aiden's house was downhill and the speed of the wagon tripled. What happened to the kinetic energy when the mass tripled? What happened to the kinetic energy when the speed tripled?

   A   When the mass tripled, the kinetic energy increased by a factor of 3.
       When the speed tripled, the kinetic energy increased by a factor of 3.

   B   When the mass tripled, the kinetic energy increased by a factor of 3.
       When the speed tripled, the kinetic energy increased by a factor of 9.

   C   When the mass tripled, the kinetic energy increased by a factor of 9.
       When the speed tripled, the kinetic energy increased by a factor of 3.

   D   When the mass tripled, the kinetic energy increased by a factor of 9.
       When the speed tripled, the kinetic energy increased by a factor of 9.

The figure below shows mass and relative kinetic energy in energy bars for three vehicles.

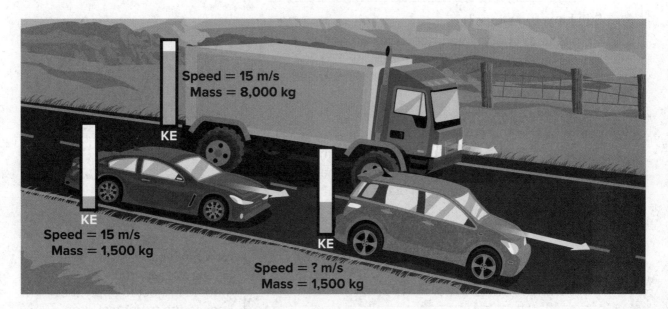

3. What can you determine about the speed of the blue car?

   A   The blue car's speed is the same as the red car's speed.

   B   The blue car's speed is less than the truck's speed.

   C   The blue car's speed is equal to the truck's speed.

   D   The blue car's speed is greater than the red car's speed.

Real-World Connection

4. **Analyze** Which vehicle will have more kinetic energy, a parked semi-truck or a car moving at 50 km/h? Explain your reasoning.

_____

_____

_____

_____

_____

_____

_____

_____

_____

_____

_____

 Still have questions?
Go online to check your understanding about kinetic energy.

REVISIT
PAGE KEELEY SCIENCE PROBES

Do you still agree with the student you chose at the beginning of the lesson? Return to the Science Probe at the beginning of the lesson. Explain why you agree or disagree with that student now.

EXPLAIN THE PHENOMENON

Revisit your claim about what determines the distance a ball travels. Review the evidence you collected. Explain how your evidence supports your claim.

START PLANNING
STEM Module Project
Science Challenge

Now that you learned about kinetic energy, go to your Module Project to apply that knowledge to vertical-drop rides. Think about how the constraints of the ride determine its kinetic energy levels.

124 EVALUATE Module: Mechanical Energy

LESSON 2 LAUNCH

Don't fall!

Yoko and her friends are learning about energy in class. They see a ball sitting on a table. Yoko and her friends have different ideas about the energy of the ball. This is what they said:

Yoko: I think the ball has no energy because it is not moving.

Caleb: I think the ball has no energy because no forces are acting on it.

Kawasi: I think the ball has energy but less energy than if it was sitting on the ground because it is farther away from Earth's surface.

Sonia: I think the ball has more energy than if it was sitting on the ground because it is farther away from Earth's surface.

Who do you most agree with? _____ Explain your thinking about how energy relates to the ball.

_____
_____
_____
_____
_____

You will revisit your response to the Science Probe at the end of the lesson.

LESSON 2
Potential Energy

ENCOUNTER THE PHENOMENON

How does changing the position of a ball affect its energy?

How could you increase the potential amount of energy of a ball? Your teacher will provide you with some materials to use. Examine the materials. Use the materials to change the position of the ball to increase the ball's potential amount of energy. Describe or illustrate what you did to increase the potential amount of energy below.

GO ONLINE
Check out *Increasing Potential* to see this phenomenon in action.

ENGAGE Lesson 2 Potential Energy 127

EXPLAIN
THE PHENOMENON

What does it mean for an object to have "potential?" If you have potential, it means you have the ability to do something. If you have the ability to do something, you must have the energy to do that task. Could you make the ball in the activity have the ability to do something? Using the ball from the activity as an example, make a claim about how you could increase the potential energy of an object.

CLAIM
You can increase the potential energy of an object, such as a ball, by....

 COLLECT EVIDENCE as you work through the lesson. Then return to these pages to record your evidence.

EVIDENCE
A. What evidence have you discovered to explain how changing the shape of an elastic object increases potential energy?

Module: Mechanical Energy

MORE EVIDENCE

B. What evidence have you discovered to explain how changing the distance between an object and Earth's surface increases potential energy?

When you are finished with the lesson, review your evidence. If necessary, based on the evidence, revise your claim.

REVISED CLAIM

You can increase the potential energy of an object, such as a ball, by....

Finally, explain your reasoning for how and why your evidence supports your claim.

REASONING

The evidence I collected supports my claim because...

LESSON 2 Potential Energy

What role does energy play when an object is not moving?

You have learned that when an object is in motion is has kinetic energy. Kinetic energy is due to the mass of the object and its speed. Take a look at the cup in the figure on the right. Does the cup have any energy when it is at rest? Let's find out!

LAB Slingshot Physics

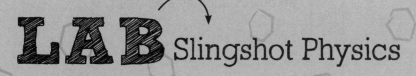

Safety

Materials
rubber band   nickel   meterstick

Procedure

1. Read and complete a lab safety form.

2. Using two fingers, carefully stretch a rubber band on a table until it has no slack.

3. Place a nickel on the table, slightly touching the midpoint of the rubber band.

4. Push the nickel back 0.5 cm into the rubber band and release. Measure the distance that the nickel travels. Record this distance in the Data and Observations section below.

5. Repeat steps 3 and 4 two more times, each time pushing the nickel back an additional 0.5 cm.

6. Follow your teacher's instructions for proper cleanup.

Data and Observations

130 EXPLORE/EXPLAIN Module: Mechanical Energy

Analyze and Conclude

7. Describe how the distance that the nickel traveled depended on the distance that you stretched the rubber band.

   _____

   _____

8. Infer how the initial speed of the nickel depended on the distance that you stretched the rubber band.

   _____

   _____

   _____

9. What happened to the energy that the rubber band contained as you increased the distance that you stretched the rubber band? Explain.

   _____

   _____

   _____

Potential Energy If the rubber band was not stretched back and released the nickel would not have moved. Somehow the rubber band gave energy to the nickel so the nickel could move. To be able to pass energy to the nickel, the rubber band must have contained energy when it was stretched. The energy contained in the rubber band is called potential energy. **Potential energy** is the energy due to interactions between objects or particles when distance changes. The amount of potential energy an object has depends on the positions of objects or particles.

When you stretch a rubber band, you are increasing a form of potential energy called elastic (ih LAS tik) potential energy. **Elastic potential energy** is energy stored in objects that are compressed or stretched, such as springs and rubber bands.

 Want more information?
Go online to read more about the types of potential energy.

FOLDABLES
Go to the Foldables® library to make a Foldable® that will help you take notes while reading this lesson.

EXPLORE/EXPLAIN Lesson 2 Potential Energy **131**

Springs When you stretch a spring like a rubber band, or compress it, you change its shape. This can be seen in the figure on the right. The spring will always try to return to its original resting, or static, shape. The more you change its shape, the greater the elastic potential energy the spring will contain.

Systems and Energy The rubber band and nickel in the *Slingshot Physics* lab represent a system. Recall there are two types of systems, open and closed. If the focus is just on the rubber band and the nickel, it is considered a closed system. When the environment surrounding the objects is involved, the system is now an open system. When looking at the energy in a system it is important to identify the components of the system, and how those components change. Did you notice that the farther back the rubber band was stretched the farther the nickel moved? The rubber band contained more potential energy as it was stretched farther back. The larger the change in position between the objects, the greater the potential energy.

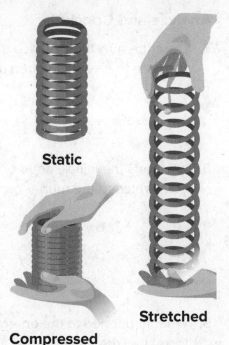

Static

Compressed

Stretched

THREE-DIMENSIONAL THINKING

Sketch **models** of the rubber band-nickel **systems** from the Lab *Slingshot Physics*. In the models, indicate the changing level of potential **energy**.

Before Rubber Band Stretch	At Maximum Rubber Band Stretch

Types of Potential Energy There are many different kinds of potential energy. These different forms involve different kinds of forces and different kinds of objects. Some types of potential energy are due to fields created by objects. These fields are invisible to the eye and extend into the space surrounding the object. Gravitational potential energy, electric potential energy, and magnetic potential energy are examples of potential energy due to forces in fields.

132 EXPLORE/EXPLAIN Module: Mechanical Energy

Determining Potential Energy There are many different equations for determining potential energy based on the forces and objects involved. What is usually important is the change in potential energy of the objects, not the potential energy of an object at a particular point in time.

COLLECT EVIDENCE

How does changing the shape of an elastic object help explain how you can increase the potential energy of a ball? Record your evidence (A) in the chart at the beginning of the lesson.

How does the distance between an object and Earth's surface affect the potential energy of the object?

If you throw a ball up in the air, you know that the ball will fall back down. The ball falls because it is in Earth's gravitational field and there is gravitational attraction between the ball and Earth. Once it falls, it is moving and has kinetic energy. Did the ball have energy before it fell?

INVESTIGATION

Dropping the Ball

A group of students completed an investigation to collect data about how much potential energy an object has as it moves farther and farther away from Earth's surface.

A summary of their procedure is recorded below.

Step 1: Measure the mass of a marble. Record the mass.

Step 2: Drop a marble from 0.5 m onto clay.

Step 3: Measure the volume of the crater made by the marble by using water drops from a pipette. This will be the indicator of the amount of potential energy contained by the marble.

Step 4: Repeat two more times from the same height.

Step 5: Repeat steps 3–4 at heights of 1.0 m, 1.5 m, and 2.0 m.

EXPLORE/EXPLAIN Lesson 2 Potential Energy **133**

1. Examine the procedure. What could the group have done to minimize error when collecting data?

The following data was collected from the experiment:

Marble 1 Mass = 7 g Diameter = 1.2 cm	Trial	Height			
		0.5 m	1.0 m	1.5 m	2.0 m
	1	2 drops	4 drops	5 drops	8 drops
	2	3 drops	4 drops	6 drops	8 drops
	3	2 drops	4 drops	6 drops	6 drops

2. Analyze the data. What cause-and-effect patterns do you notice?

3. What conclusion can you draw from the data about the amount of energy the marble has at different heights? Explain.

Another group followed the same procedure but they used a different marble. The following data was collected from the experiment:

Marble 2 Mass = 2.5 g Diameter = 1.2 cm	Trial	Height			
		0.5 m	1.0 m	1.5 m	2.0 m
	1	1 drop	1.5 drops	2 drops	3 drops
	2	1 drop	2 drops	4 drops	2 drops
	3	1 drop	2 drops	3 drops	3 drops

4. Compare the data to the first data table. What are some similarities and differences between the data sets?

134 EXPLORE/EXPLAIN Module: Mechanical Energy

5. How were the two marbles similar? How were they different?

_____

_____

6. What conclusion can you draw from the data about the amount of energy and the mass of the marble?

_____

_____

Gravitational Potential Energy You have seen that an object can have energy even if it is not moving. If you hold a ball 1 m above the ground and let it go, you know that the ball will fall. The ball falls because the ball is in Earth's gravitational field. There is gravitational attraction between the ball and Earth. To increase the energy of an object in relation to Earth's gravitational field a force must be applied to move these two objects farther apart. This type of energy is called gravitational potential energy. **Gravitational potential energy** is stored energy due to the interactions of objects in a gravitational field. Gravitational potential energy exists between all objects but is only significant when at least one very large object is involved, such as Earth.

Recall that objects can be located anywhere inside a gravitational field. The farther away the object is from the center of the gravitational field, the more gravitational potential energy the object has in relation to the field. The difference in gravitational potential energy depends on the change in position of the object and the object's mass.

Change in Position Examine the figure of the girl and the backpack on the right. The girl in the figure increases the gravitational potential energy between her backpack and Earth by lifting the backpack. To increase the gravitational potential energy of the backpack, she exerts a force on the backpack that opposes the force of gravity. When she applies this force, she increases the energy of the backpack. The potential energy of the backpack increases as its height increases.

Mass What if the girl removed some of the books from her backpack? She would exert less force to pick up the backpack because it would have less mass. Less energy would be needed to lift the backpack to the same height. Gravitational potential energy and mass have a proportional relationship. The potential energy of the backpack decreases as its mass decreases.

THREE-DIMENSIONAL THINKING

Sketch a **model** to show that when the distance between objects changes the gravitational potential **energy** of the **system** changes.

Explain how the model shows what happens to the gravitational potential energy of the system when the distance between the objects increases.

COLLECT EVIDENCE

How does changing the distance between an object and Earth's surface help explain how you can increase the potential energy of a ball? Record your evidence (B) in the chart at the beginning of the lesson.

136 EXPLORE/EXPLAIN Module: Mechanical Energy

STEM Careers

A Day in the Life of a Roller Coaster Designer

Riding roller coasters in amusement parks can give you the feeling of danger, but these rides are designed to be safe. Engineers use the laws of physics to design amusement park rides that are thrilling but harmless. Engineers use the concepts of gravitational potential energy to design the drops and hills found in roller coasters.

The first thing to do when designing a roller coaster is to pick what it will be made of. Roller coasters are constructed of steel or wood. The next thing to think about is what kind of ride the experience is going to be. The roller coaster can be death-defyingly tall or just tall enough to get the speeds needed to go through any set of loops and turns. The design can be traditional or suspended, and have lots of loops or just be straight. It can be a long or short ride.

To design a roller coaster, the designer must be creative but also have a good sense of algebra and geometry. She must understand the role of energy and forces on the coaster.

It's Your Turn

ENGINEERING Connection Research some designs of roller coasters. Use these designs to create your own track for a roller coaster as a sketch or a model. Share your design with your class to gain feedback on how you might be able to improve your design. Incorporate the feedback into your roller coaster.

LESSON 2
Review

Summarize It!

1. **Model** Create a model to illustrate the interactions that affect the amount of potential energy an object might have.

Three-Dimensional Thinking

Use the diagram below to answer questions 2–3.

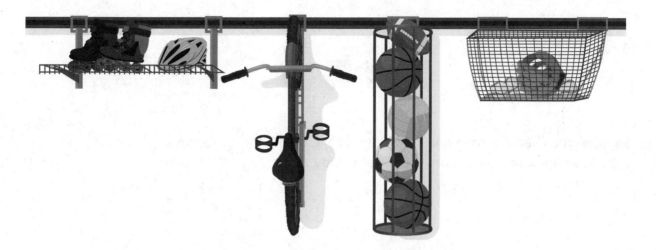

2. In the rack there are two basketballs. Which basketball has more energy?

   A The top basketball has more energy because it is farther away from surface of Earth.

   B The bottom basketball has more energy because it is closer to surface of Earth.

   C Both basketballs have the same amount of energy because they have the same mass.

   D Both basketballs have the same amount of energy because they are not moving.

3. In the rack there is a bike helmet and a pair of roller skates. The mass of the bike helmet is 550 g and the mass of a roller skate is 2,000 g. Which item has more energy?

   A The bike helmet has more energy because it has less mass.

   B The roller skate has more energy because it is has more mass.

   C Both items have the same amount of energy because they are the same distance from the surface of Earth.

   D Both items have the same amount of energy because they are not moving.

EVALUATE Lesson 2 Potential Energy 139

Real-World Connection

4. **Infer** How could you increase the gravitational potential energy between yourself and Earth?

_____

_____

_____

_____

5. **Explain** You overhear someone say that two objects with the same mass must have the same gravitational potential energy. Do you agree? Explain.

_____

_____

_____

_____

_____

_____

 Still have questions?
Go online to check your understanding about the types of potential energy.

 **REVISIT SCIENCE PROBES** Do you still agree with the student you chose at the beginning of the lesson? Return to the Science Probe at the beginning of the lesson. Explain why you agree or disagree with that student now.

KEEP PLANNING
STEM Module Project Science Challenge

Now that you've learned about potential energy, go back to your Module Project to start developing your model of the vertical-drop ride. Be sure to keep in mind how energy changes as amusement park rides change riders' positions.

 **EXPLAIN THE PHENOMENON** Revisit your claim on how you can increase the potential energy of the ball. Review the evidence you collected. Explain how your evidence supports your claim.

LESSON 3 LAUNCH

Swing Low

Manuel was watching his younger siblings on the swing set. He had just learned about kinetic and potential energy in school. He wondered about energy's role in the back-and-forth movement of a swing. Which statement do you think explains the energy of the swing best?

A. When a person is at the highest point of a swing, there is no energy, because he is not moving.

B. A person has the most energy at the lowest point of the swing, because there he is moving fastest.

C. The total amount of energy remains the same. It just changes from one form to another as the person swings back-and-forth.

Explain your thinking about the energy of the swing.

_____
_____
_____
_____
_____

You will revisit your response to the Science Probe at the end of the lesson.

LESSON 3
Conservation of Energy

142 ENGAGE Module: Mechanical Energy

ENCOUNTER THE PHENOMENON

What happens to the energy of the weight as it swings?

GO ONLINE

Watch the video *Pendulum Demo* to see this phenomenon in action.

Watch the video. Describe the energy of the weight when it is A) pulled back to the farthest extent, B) at the bottom of the swing, C) at the highest point on the other side, and D) back at the original side.

A	B

C	D

ENGAGE Lesson 3 Conservation of Energy **143**

EXPLAIN
THE PHENOMENON

The weight swung in an arc away and back again. What ideas do you have about energy's role as the weight swings? Does the weight gain or lose energy? Does the energy change? Use the idea of a swinging weight to make a claim about the energy an object has as it changes position.

CLAIM
When an object, such as a pendulum, changes position, its energy...

COLLECT EVIDENCE as you work through the lesson. Then return to these pages to record your evidence.

EVIDENCE
A. What evidence have you discovered to explain the role of kinetic energy and potential energy in the motion of a pendulum?

MORE EVIDENCE

B. What evidence have you discovered to explain how energy is transferred into or out of a system?

When you are finished with the lesson, review your evidence. If necessary, based on the evidence, revise your claim.

REVISED CLAIM

When an object, such as a pendulum, changes position, its energy...

Finally, explain your reasoning for how and why your evidence supports your claim.

REASONING

The evidence I collected supports my claim because...

LESSON 3 Conservation of Energy

What types of energy does an object have if it is both moving and off the ground?

Kinetic energy and potential energy are two classifications of energy. Take a look at the water slide on the right. As the car slides down, what type of energy does it have? Can it have more than one type of energy? Let's find out!

What energy does the ride have?

> **Want more information?**
> Go online to read more about conservation of energy in systems.

> **FOLDABLES**
> Go to the Foldables® library to make a Foldable® that will help you take notes while reading this lesson.

LAB: The Energy of a Pendulum

Safety

Materials

ring   washers (5)
string, 20 cm   meterstick
ring stand   stopwatch

Procedure

1. Read and complete a lab safety form.

2. Set up the ring and ring stand. Use the meterstick to adjust the ring to a height of 35 cm above the table or desk.

3. Securely tie the string to the 5 washers. Measure 15 cm of length of the string. Tie the string at this point to the ring. You have created a pendulum.

4. Allow the pendulum to hang at rest. Consider the energy of the pendulum. On the lines below, record the type of energy you think the pendulum has while hanging at rest. Explain.

_____

_____

146 EXPLORE/EXPLAIN Module: Mechanical Energy

5. Hold the pendulum above the table to form a small angle with the ring stand. Record the type of energy you think the pendulum has when it is held at an angle. Explain.

_____

_____

6. Record the height of the pendulum at the raised position. From the raised position, release the pendulum and allow it to swing. Use the meterstick to record the maximum height the pendulum reaches on the other side of the ring. Record the heights and the motion of the pendulum in the Data and Observations section below.

7. Allow the pendulum to swing for 1 minute. Record your observations.

8. Follow your teacher's instructions for proper cleanup.

Data and Observations

Analyze and Conclude

9. What force acted on the pendulum when it was released from its raised position?

_____

_____

10. Based on your observations, when did the pendulum have kinetic energy? Explain.

_____

_____

11. Based on your observations, when did the pendulum have potential energy? Explain.

_____

_____

EXPLORE/EXPLAIN Lesson 3 Conservation of Energy

Analyze and Conclude, continued

12. On the model below, label the types of energy the pendulum has at each part of the arc. Also label if that energy is at its maximum, its minimum, is increasing, or is decreasing.

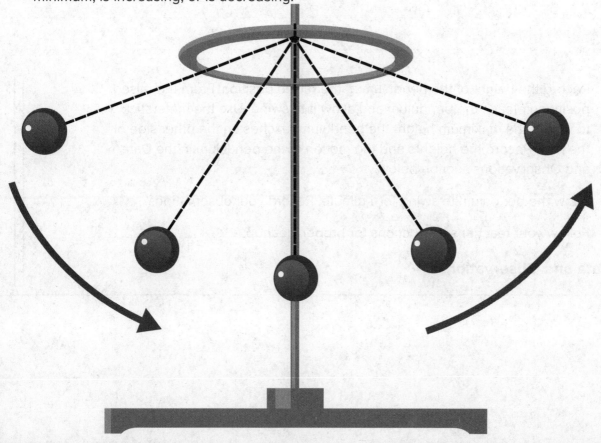

Mechanical Energy When the pendulum was swinging in the lab, did it have kinetic energy? It had mass and speed, so it had kinetic energy. The pendulum was also above its resting height, so it had gravitational potential energy. The pendulum had both kinetic and potential energy. Sometimes it is useful to examine the individual energy types. However, it is also important to consider the energy of the entire system. The sum of the potential energy and the kinetic energy in a system is **mechanical energy.** All forms of energy are measured in joules (J).

GO ONLINE for additional opportunities to explore!

Investigate mechanical energy by performing one of the following activities.

☐ **Model** the energy of a skateboarder in the **PhET Interactive Simulation** *Energy Skate Park: Basics.*

OR

☐ **Explain** the energy of a roller coaster after watching the **Animation** *Energy Transfers.*

Changes Between Kinetic and Potential Energy In many cases, during the change of motion of an object, kinetic energy changes to potential energy and vice versa. Think about tossing a basketball in the air. The ball is moving fastest and has the most kinetic energy as it leaves your hand. As the ball moves upward, its speed and kinetic energy decrease. However, the potential energy is increasing because the ball's height is increasing. Kinetic energy is changing into potential energy. At the ball's highest point, the gravitational potential energy is at its greatest, and the ball's kinetic energy is at its lowest.

As the ball moves downward, its potential energy decreases. At the same time, the ball's speed increases. Therefore, the ball's kinetic energy increases. Potential energy is transferred into kinetic energy. Immediately before the ball reaches the other player's hand, its kinetic energy is back to the maximum value.

Did you notice that as the kinetic energy of the basketball decreased, the potential energy increased by the same amount? The amount of kinetic energy at the beginning of the toss determined how high the basketball would go and therefore the amount of potential energy the ball could gain. On the way back down, the height the basketball reached determined the speed of the ball. This determined the amount of kinetic energy the ball had as it fell.

Think back to the pendulum in *The Energy of a Pendulum* lab. You may have noticed that the pendulum reached the same height on the other side as the height that you released it from. This is because its potential energy transferred to kinetic energy as it swung to the bottom of the arc. Then the kinetic energy transferred to potential energy as it swung back up to the other side. Energy was constantly transferred back and forth. Later in the lesson you will explore how the pendulum gains more total energy and what happens to the total energy over time.

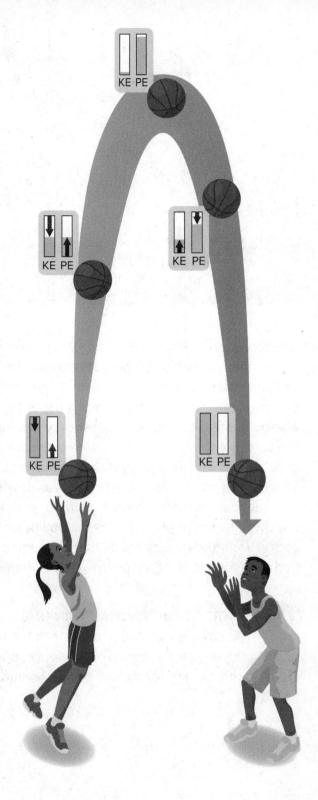

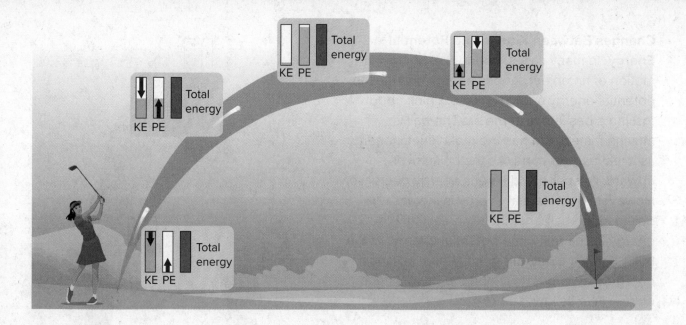

Conservation of Energy Examine the path of the golf ball in the figure above. Energy bars are shown at each point of the golf ball's motion. As the ball separates from the club, kinetic energy is the greatest and potential energy is the least. The total energy bar is full. As the ball moves up, kinetic energy decreases and potential energy increases. The total energy remains the same. At the highest point, where gravitational potential energy is the greatest and kinetic energy is the least, the total energy is still unchanged. This pattern continues as the ball moves downward. The decrease in gravitational potential energy and increase in kinetic energy do not change the overall amount of energy as the ball nears the ground. The **law of conservation of energy** states that although energy is always transferring from one form to another, energy is not created or destroyed.

THREE-DIMENSIONAL THINKING
Recall the pendulum from the Lab *The Energy of a Pendulum*. **Construct an argument,** based on evidence, on whether the pendulum followed the pattern for the law of conservation of energy.

COLLECT EVIDENCE

How do both kinetic energy and potential energy help explain the motion of the pendulum? Record your evidence (A) in the chart at the beginning of the lesson.

How does energy transfer into or out of a system?

You just discovered that an object can have both potential and kinetic energy. As the object moves, kinetic energy and potential energy can transfer from one form to the other without being lost. Think about the pendulum from *The Energy of a Pendulum* lab. How did the pendulum get the energy to start swinging? If you want to increase the energy of an object, where does that energy come from? Let's find out!

LAB So Much Work

Safety

Materials

string, 1 m   paper clip   tape
meterstick   small box   metric ruler
washers (5)

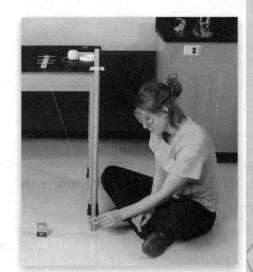

Procedure

1. Read and complete a lab safety form.

2. Use the photo on the right as a guide to make a pendulum. Hang the washers on the paper clip. Place the box so it will block the swinging pendulum. Mark the position of the box with tape.

3. Pull the pendulum back until the bottom of the washers are 15 cm from the floor. Release the pendulum. Measure and record the distance the box moves in the Data and Observations section below. Repeat two more times.

4. Follow your teacher's instructions for proper cleanup.

Data and Observations

	Trial 1	Trial 2	Trial 3
Distance Box Moves			

EXPLORE/EXPLAIN Lesson 3 Conservation of Energy

Analyze and Conclude

5. What signs indicated that an energy transfer happened to both the pendulum and to the box?

_____

_____

6. Use the energy flow diagram to model the components of the system at the moment the pendulum hit the box. Identify the type of energy involved and whether the energy increased or decreased.

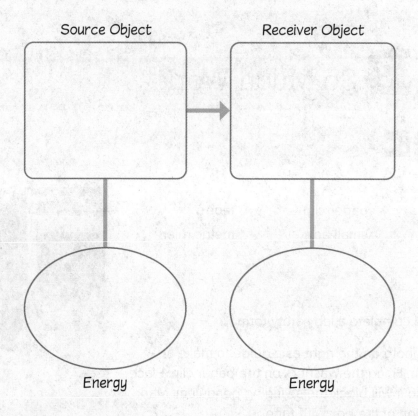

7. What caused the box to move?

_____

_____

_____

Energy Transfer Signs A change in an object's motion is an indication that an energy transfer occurred between objects. You may have heard a sound as the pendulum collided with the box. Some of the mechanical energy was used to create a vibration which was heard as sound. A change in the sound of a system is also an indication that an energy transfer occurred.

Transferring Energy The energy of a system can change only if energy is transferred to or from the system. One way to transfer energy to a system is to do work on an object. **Work** is the transfer of energy to an object by a force that makes an object move in the direction of the force. Work is only being done while the force is applied to the object. You pulled back on the pendulum with a force. In other words, you did work to the pendulum. This increased the energy of the pendulum. When you released the pendulum, it swung back down and hit the box. The pendulum applied a force to the box causing the box to gain energy and move.

THREE-DIMENSIONAL THINKING

Examine the system model below. Construct an explanation for why the amount of energy in the system of the box and shelf increased.

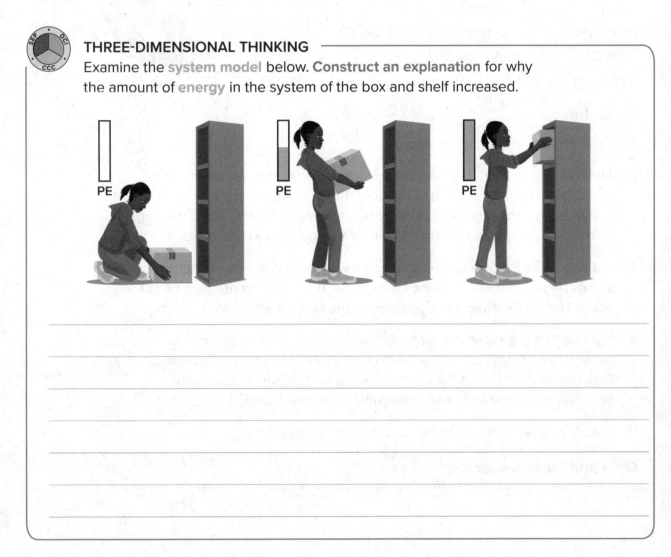

Doing Work Imagine pushing your bicycle into a bike rack. Your push (a force) makes your bicycle (an object) move. Therefore, work is done. You do no work if you push on the bicycle and it does not move. A force that does not make the object move does no work. Suppose you let go of the bicycle and it continues moving forward. Even though the bicycle is still moving, no work is being done. This is because you are no longer applying a force to the bicycle.

An object can gain energy when a force does work on the object. What causes energy to be transferred out of a system? Let's find out!

LAB Double Pendulum

Safety

Materials

string
ring stands (2)
rings (2)
meterstick
scissors
washers of equal mass (2)
stopwatch

Procedure

1. Read and complete a lab safety form.

2. Attach a ring to each ring stand about one-fourth of the way down from the top. Place the two ring stands about 30 cm apart. Tie a piece of string to the ring on one ring stand. Attach the other end of the string to the ring on the other ring stand. There should not be any slack in the string.

3. Cut two 15-cm lengths of string. Attach a washer to one end of each string. Tie the other ends of the string to the horizontal string, about 5 cm apart. The washers must hang at exactly the same height.

4. Pull one of the washers upward and backward in the direction of the nearest ring stand. Pull the washer to the height of the horizontal string. Release the washer. Let the washers swing for 1 minute. Record your observations in the Data and Observations section below.

5. Follow your teacher's instructions for proper cleanup.

Data and Observations

Analyze and Conclude

6. What had to happen to get the first washer to move?

7. The two pendulums will eventually slow down and stop if you wait long enough. What energy transfers take place to stop the pendulums?

_____

_____

Thermal Energy Transformations Did you notice in the Lab *Double Pendulum* that the pendulums slowed down or stopped? The pendulums exerted forces on each other, causing a change in motion to happen. This indicates that there was a transfer of energy. But what caused the change in energy? Were the pendulums in contact with any other forces?

Recall there is always some friction between any two surfaces that are rubbing against each other. In this case, friction in the form of air resistance acted on the pendulums. As a result, some mechanical energy was transferred to the air as thermal energy when the two surfaces (air and the pendulum) rubbed against each other. Changes in thermal energy are shown by changes in temperature.

You probably didn't notice an increase in the temperature of the air around the pendulums. This is because only a small amount of energy was transferred over time. A more obvious example of mechanical energy transferring to thermal energy due to friction is when a tire leaves tire marks on the road. In this case, a large amount of energy is transferred to thermal energy due to the force of friction between the tires and the road. The thermal energy causes the surface of the tires to melt slightly. This leaves tire marks on the road. A change in temperature is another indication that an energy transfer has occurred.

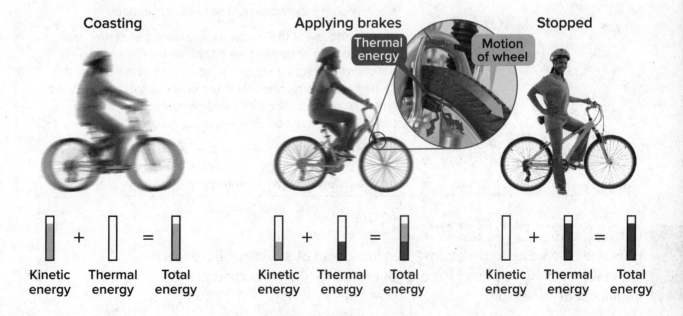

EXPLORE/EXPLAIN Lesson 3 Conservation of Energy **155**

Read a Scientific Text

People often use the term "running out of energy." For example, when all the gas in a car has been burned, the car stops running. Did the car run out of energy? How does this follow the law of conservation of energy?

CLOSE READING

Inspect
Read the passage *Energy—conserved or not?*

Find Evidence
Reread the section *Running Out of Energy.* Underline how thermal energy is produced when a car's gasoline burns.

Make Connections
Communicate With your partner, discuss the two laws of thermodynamics. Does the second law of thermodynamics contradict the first law?

Energy—conserved or not?

The law of conservation of energy states that energy is not created or destroyed but can change from one form to another. This law is also called the first law of thermodynamics. Thermodynamics is the study of the conversion of energy into heat and other forms.

Running Out of Energy To understand the first law of thermodynamics, you have to focus on the energy exchanges in a particular system. A system is the part of the universe involved in the changes you are observing. For example, think about the car and gasoline system. When gasoline burns, the energy stored in the gasoline changes to energy that moves parts of the car. As the parts move, friction converts the moving energy to thermal energy. Molecules in the air around the parts then move faster.

Although the quantity of energy doesn't change, the quality of the energy will. The potential energy in the gasoline is useful energy because it can be used to do work on other objects. But as other forms of energy change to thermal energy, the energy can no longer be used to do useful work.

Entropy The second law of thermodynamics is called the law of entropy. Entropy is a measure of the disorder or randomness of a system. Gasoline molecules have low entropy because there is a high amount of energy in a small system. The heated molecules in air that are the final product of the burning of gasoline are more randomly spread out and so have high entropy.

The second law of thermodynamics states that matter and energy tend to become more disordered. Think about a cup of hot cocoa cooling off at room temperature. In the cocoa, the energy is concentrated only in the molecules of liquid. By the time the cocoa cools off, that energy has spread throughout the room. Entropy—disorder—is much greater.

Because of the second law, we can say that the universe will run out of useful energy when all the energy is spread out with maximum randomness, or entropy.

COLLECT EVIDENCE

How does the way energy is transferred into or out of a system help explain the motion of the pendulum? Record your evidence (B) in the chart at the beginning of the lesson.

A Closer Look: Creating Electrical Energy

If you have ever stood underneath a shower, you have felt the energy of falling water. Falling water is the most widely used renewable energy resource. Hydroelectric power plants generate electric energy from falling water.

First a large source of potential energy needs to be created. A dam is built across a flowing water source such as a river. The water behind the dam builds up and is at a higher level than the water below the dam. As water falls through tunnels in the dam, the water's potential energy becomes kinetic energy. The moving water is channeled through a turbine. The kinetic energy of the moving water does work on the blades of the turbine, causing them to spin. As the turbine spins, the mechanical energy produced is used to make electrical energy in a generator. The electrical energy is then carried by wires to transformers and then to your home!

It's Your Turn

Research Water is not the only resource used to generate electric energy. What other processes are used to produce electrical energy? Research and then create a diagram showing one of the processes and the energy transfers that happen to produce electrical energy. Indicate where forces do work. Present your diagram to the class.

LESSON 3
Review

Summarize It!

1. **Describe** how energy changes as a ball is thrown in the air. Compare the kinetic and potential energy of the ball at each position shown in the model of the system shown below.

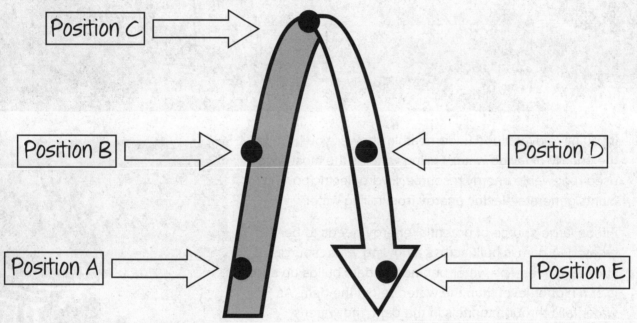

Position	Kinetic Energy	Potential Energy
A		
B		
C		
D		
E		

Three-Dimensional Thinking

The graph shows the kinetic energy (KE) and potential energy (PE) of a bouncing ball over a period of 9 seconds. Use the graph to answer questions 2–3.

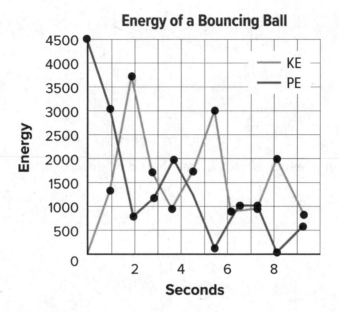

2. When does the ball have the most kinetic energy?

   A At the start, because it is the highest off the ground.

   B At the end, because it is moving fastest as it approaches the ground for the final bounce.

   C At second 2, because it is moving fastest as it approaches the ground for the first bounce.

   D At second 4, because it has reached its second highest bounce.

3. What statement best describes what happens to the total mechanical energy of the ball as it bounces?

   A The total amount of energy remains the same because the law of conservation of energy says energy is conserved.

   B The total amount of energy goes down because some energy becomes thermal energy due to the force of friction acting on the ball.

   C The total amount of energy goes up because energy is gained from the force of gravity every time the ball bounces.

   D The total amount of energy remains the same because work is being done on the ball by the force of gravity every time the ball bounces.

Real-World Connection

4. **Model** Think of a situation that you are familiar with where the gravitational potential energy and kinetic energy of an object are constantly shifting back and forth. Create a model of the energy transfers that occur for the object.

5. **Construct an argument** about whether energy is transferred when you apply the brakes on a bicycle.

___

 Still have questions?
Go online to check your understanding about conservation of energy in large systems.

 Do you still agree with the statement you chose at the beginning of the lesson? Return to the Science Probe at the beginning of the lesson. Explain why you agree or disagree with that statement now.

 Revisit your claim on how the energy of the pendulum changes as it swings back and forth. Review the evidence you collected. Explain how your evidence supports your claim.

FINISH PLANNING AND PRESENT
STEM Module Project Science Challenge

Now that you've learned about mechanical energy transfers, go back to your Module Project to finalize your model and start constructing your argument. Keep in mind how energy can transfer on amusement park rides.

STEM Module Project
Science Challenge

Energy at the Amusement Park

The CEO of the local amusement park has heard that energy is lost during a run of the park's vertical-drop ride. He wonders if they could collect the energy to power the park's generator. Your company has been contracted to develop a model detailing the change in position, speed, forces, and types of energy during a run of the ride. With your team, research different vertical-drop rides. Next, develop your model. Finally, construct an argument on whether or not energy can be transferred to power a generator for the park.

Planning After Lesson 1

Research three vertical-drop rides. Find out their names, locations, drop heights, riders per vehicle, and top speeds. Record your findings in the space below.

STEM MODULE PROJECT Module: Mechanical Energy

STEM Module Project
Science Challenge

Planning After Lesson 1, continued

Based on the factor of top speed, sequence the rides from highest to lowest amounts of kinetic energy assuming mass is the same for all rides. Explain your order.

The average adult human weighs 62 kg. Based on the factor of mass, sequence the rides from highest to lowest amounts of kinetic energy assuming maximum speed is the same for all rides. Assume maximum number of riders per vehicle. Explain your order.

Planning After Lesson 2

Return to your research from Lesson 1. Based on the factor of drop height, sequence the rides from highest to lowest amounts of potential energy assuming mass is the same for all rides. Explain your order.

Develop a model of a vertical-drop ride to describe how the arrangement of objects interacting at a distance affects the amount of potential energy stored in the system. For your model, be sure to identify:

- the system of objects,
- the force through which the objects interact, and
- the distance between the objects and its effect on the potential energy.

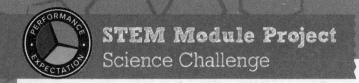

Planning After Lesson 3

Revise the model of your vertical-drop ride to include an analysis of the change in position, speed, forces, and types of energy during a run of the ride. Include the following:

- locations where changes in kinetic energy occur,
- location where changes in gravitational potential energy occur,
- locations where forces are used to input energy into the ride, and
- locations where energy leaves the ride.

Draw or describe your revised model in the space below.

Construct Your Argument

Construct the argument that you will share with the CEO of the amusement park about whether energy can be transferred from the vertical-drop ride to power a generator for the park.

Identify your argument's claim.

What information and evidence support your claim?

Prepare for your audience to question your sources by stating what you did to ensure that the evidence and information used to support your claim is accurate and reliable and identify any alternative interpretations of the evidence.

STEM Module Project
Science Challenge

Construct Your Argument, continued

Use your answers to the previous questions to develop your argument. Support your argument with your model. Be sure that your reasoning is clear. Record your argument in the space below.

Create Your Presentation

Develop a presentation, integrating multimedia and visual displays, with your argument for the amusement park CEO on whether energy from the vertical-drop ride could be transferred to a generator. Present your argument.

Congratulations! You've completed the Science Challenge requirements!

Module Wrap-Up

REVISIT THE PHENOMENON

Think about everything you have learned in the module about how mechanical energy governs the motions of objects. Construct an explanation for the role energy plays on an amusement park ride.

OPEN INQUIRY

What are one or two questions you still have about the phenomenon?

_____
_____
_____
_____

Choose the question that interests you the most. Plan and conduct an investigation to answer this question.

Electromagnetic Forces

ENCOUNTER
THE PHENOMENON

How is a battery-powered fan similar to this hand-cranked flashlight?

Power on!

GO ONLINE
Watch the video *Power on!* to see this phenomenon in action.

Collaborate Think about what you know about how the fan and the flashlight work. With a partner, discuss his or her thoughts. Then record what you both would like to share with the class below.

Module: Electromagnetic Forces

STEM Module Project Launch
Engineering Challenge

The Great Metal Pick-Up Machine

Lesson 1 Magnetic Forces

Lesson 2 Electric Forces

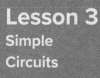

Lesson 3 Simple Circuits

Lesson 4 Electromagnetism

Your friend's younger brother, Andy, loves his sandbox and all types of construction machinery. His favorite machine is the big crane that can pick up a lot of scrap metal at once. He thinks they are so cool!

Your friend has asked for your help to build Andy his very own crane that he can use in his sandbox to pick up his metal toy cars. First, you will need to determine which type of forces the crane might use. You will then design and test your crane. Using the data from each test, you will optimize your crane design before presenting the crane to your friend.

Start Thinking About It

Examine the photo. Formulate questions about the factors that might affect the strength of the magnetic force. Share your questions with your group.

STEM Module Project
Planning and Completing the Engineering Challenge How will you meet this goal? The concepts you will learn throughout this module will help you plan and complete the Engineering Challenge. Just follow the prompts at the end of each lesson!

170 STEM MODULE PROJECT Module: Electromagnetic Forces

LESSON 1 LAUNCH

Which pole is it?

The top picture shows a bar magnet with the poles labeled N and S. The magnet was cut into two pieces—one short piece and one long piece. How should the cut end of the short piece of magnet be labeled? Circle the answer that best matches your thinking.

A. N

B. S

C. No label—it no longer has a N or S pole on the cut end.

Explain your thinking. What rule or reasoning did you use to decide how to label the cut end of the short piece of magnet?

_____
_____
_____
_____
_____
_____

You will revisit your response to the Science Probe at the end of the lesson.

LESSON 1
Magnetic Forces

ENCOUNTER THE PHENOMENON | Why doesn't the plant fall down?

Did you notice that the plant is hovering above the table in the photo on the left? Try your own hand at getting a paper clip to hover. Attach one end of a string to a paper clip. Attach the other end of the string to the table. Move a magnet around the paper clip. Record your observations in the space below.

GO ONLINE
Watch the video *Magnet Magic* to see this phenomenon in action.

EXPLAIN
THE PHENOMENON

Did you notice that you could use the magnet to get the paper clip to move without touching it? How was this similar to the floating plant? A change in motion or preventing an object from falling can only happen when a force is acting on an object. What force was acting on the paper clip? What about the forces acting on the plant? Make a claim on how magnets can keep a plant from falling.

CLAIM
Magnets can keep a plant from falling because...

 COLLECT EVIDENCE as you work through the lesson. Then return to these pages to record your evidence.

EVIDENCE

A. What evidence have you discovered to explain the existence of magnetic forces?

B. What evidence have you discovered to explain the existence of magnetic fields?

174 Module: Electromagnetic Forces

MORE EVIDENCE

C. What evidence have you discovered to explain magnetic potential energy?

D. What evidence have you discovered to explain magnetic domains?

When you are finished with the lesson, review your evidence. If necessary, based on the evidence, revise your claim.

REVISED CLAIM
Magnets can keep a plant from falling because…

Finally, explain your reasoning for how and why your evidence supports your claim.

REASONING
The evidence I collected supports my claim because…

Lesson 1 Magnetic Forces

What is a magnetic force?

Were you able to move the paper clip without touching it? If you use magnets, you may know that magnets attract some objects, such as paper clips, but not other objects, such as pieces of paper. What other materials will act like the paper clip when near a magnet?

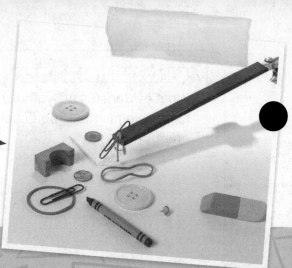

What does the magnet attract?

LAB Paper Clip Pick Up

Safety

Materials

paper clips (20) nickel
magnet wooden craft stick
penny two additional materials (you choose)

Procedure

1. Read and complete a lab safety form.

2. Touch your magnet to a pile of paper clips. Count the number of paper clips your magnet will pick up. Record your observations in a table in the Data and Observations section below.

3. Cover the ends of the magnet one at a time with a penny, a nickel, a craft stick, and two items of your choosing. Test the number of paper clips the magnet will pick up each time. Record your observations in the data table.

4. Follow your teacher's instructions for proper cleanup.

Data and Observations

Analyze and Conclude

5. What materials attracted the most paper clips? What do you think caused the attraction?

___

Magnetic Materials A **magnet** is an object that attracts iron and other materials that have magnetic qualities similar to iron. A magnet attracts paper clips and some nails because they contain iron. Any material that is strongly attracted to a magnet is a magnetic material. Magnetic materials often contain ferromagnetic (fer oh mag NEH tik) elements. Ferromagnetic elements include iron, nickel, and cobalt, which have an especially strong attraction to magnets.

A force of attraction or repulsion created by a magnet is a **magnetic force**. A magnetic force is a noncontact force which means it can apply a force without being in direct contact with another object.

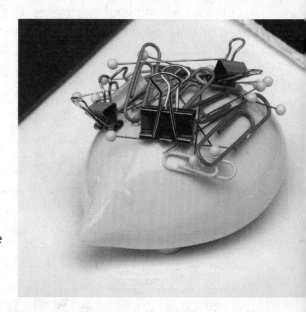

THREE-DIMENSIONAL THINKING
Analyze the data you collected from the Lab *Paper Clip Pick Up*. Develop two or more questions based on the data that you could investigate to find out more about magnetic forces. In each question underline the cause and circle the effect.

___

Want more information?
Go online to read more about magnetic fields and forces.

FOLDABLES
Go to the Foldables® library to make a Foldable® that will help you take notes while reading this lesson.

EXPLORE/EXPLAIN Lesson 1 Magnetic Forces   **177**

Where is the force of a magnet the strongest?

Why are these labeled?

You may have noticed that the ends of some magnets are different colors or are labeled *N* and *S*. What do those mean? Do they affect how many paper clips you can pick up?

Make a prediction on how the location along a bar magnet affects the number of paper clips that can be picked up. Record your hypothesis in the first column in the table below. Compare hypotheses among your classmates. List the possible effects based on your classmates' responses in the second column.

Cause	Effect

LAB The Strength of Magnets

Safety

Materials
bar magnet   paper clips (20)

Procedure
1. Read and complete a lab safety form.
2. Hold a bar magnet horizontally, and put a paper clip on one end. Touch a second paper clip to the end of the first one.
3. Continue adding paper clips until none will stick to the end of the chain. Record the number of paper clips that the magnet holds in the Data and Observations section on the next page. Remove the paper clips.

178 EXPLORE/EXPLAIN Module: Electromagnetic Forces

4. Repeat steps 2 and 3 three more times using different locations on the magnet. First, start the chain about 2 cm from the end of the magnet. Next, start the chain near the center of the magnet. Finally, start the chain at the opposite end of the magnet. Record the number of paper clips each location picks up.

5. Follow your teacher's instructions for proper cleanup.

Data and Observations

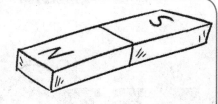

Analyze and Conclude

6. Compare the attraction at the center of the magnet with the attraction at the ends.

_____
_____
_____

7. Think back to your predictions before the lab. Evaluate each predicted outcome. Which cause-and-effect relationship most accurately describes the outcome of the lab?

_____
_____
_____
_____

8. How did the procedure help you identify cause-and-effect relationships? How could you modify the procedure to better evaluate the cause-and-effect relationships?

_____
_____
_____

Magnetic Poles The colors and labels on some magnets identify a magnet's magnetic poles. A **magnetic pole** is a place where the force a magnet applies is strongest. There are two magnetic poles on all magnets—a north pole (N) and a south pole (S). If you break a magnet into pieces, each piece will have a north pole and a south pole.

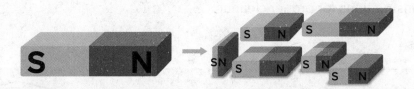

LAB Magnetic Personality

You just learned that all magnets have magnetic poles. Can you use this information to determine where the north and south poles are on an unmarked magnet?

Safety

Materials

magnets with markings (2)   magnet with no markings

Procedure

1. Read and complete a lab safety form.

2. Bring two ends that have the same markings near each other. Record your observations in the Data and Observations section on the right.

3. Bring two ends that have different markings near each other. Record your observations.

4. Bring each end of a marked magnet near one end of the unmarked magnet. Use your observations to determine what markings should be on each end of the unmarked magnet.

5. Follow your teacher's instructions for proper cleanup.

Data and Observations

180 EXPLORE/EXPLAIN Module: Electromagnetic Forces

Analyze and Conclude

6. Identify the phenomenon you were investigating.

7. How were you able to determine the markings on the unmarked magnet?

The Forces Between Magnetic Poles A force exists between the poles of any two magnets. If similar poles of two magnets, such as north and north or south and south, are brought near each other, the magnets repel. This means that the magnets will push away from each other. If the north pole of one magnet is brought near the south pole of another magnet, the two magnets attract. This means the magnets will pull together. In other words, similar poles repel, and opposite poles attract.

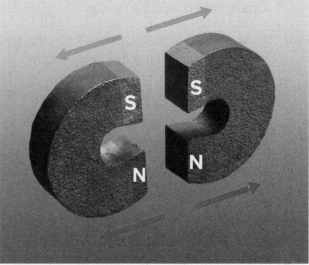

COLLECT EVIDENCE
How do magnetic forces explain why magnets can keep a plant from falling? Record your evidence (A) in the chart at the beginning of the lesson.

How can you model the force of a magnet?

The poles of the magnet are where the magnetic force is the strongest. But, that is not the only place a magnetic force exists. How can magnetic forces around a magnet be visualized? Let's investigate!

EXPLORE/EXPLAIN Lesson 1 Magnetic Forces **181**

LAB  Magnetic Fields

Safety

Materials

paper
bar magnet
compass
metric ruler

Procedure

1. Read and complete a lab safety form.

2. Step away from your magnet and all metal objects in the room. Hold the compass level in your hand. Face the direction indicated by the compass. Record the direction you are facing in the Data and Observations section below.

3. Move to two other locations. Face the direction indicated by the compass. Record the direction you are facing.

4. Bring the compass near the magnet. Record your observations.

5. On a piece of paper, draw a circle with a diameter 2 cm longer than your bar magnet. Place the bar magnet in the center of your circle. Label the circle with the magnet's north and south poles.

6. With the bar magnet at the center of the circle, place the compass on the paper just outside the circle. Draw an arrow on the circle pointing in the same direction as the north end of the compass.

7. Repeat step 6 in at least eight more positions around the circle.

8. Record which magnetic pole the north end of the compass needle points to.

9. Follow your teacher's instructions for proper cleanup.

Data and Observations

Analyze and Conclude

10. Explain the purpose of the procedure.

_____

_____

11. Why do you think the compass needle points in the direction that it does when not around a magnet?

_____

_____

12. Model the magnetic forces around a bar magnet. Use your observations of the compass needle to help build your model.

13. How does your model provide evidence for the existence of an invisible force around the magnet?

_____

_____

14. Did this lab provide evidence to support the existence of an invisible force that extends around a magnet, even affecting objects that it does not touch? How could you improve the procedure to collect more data to support this claim?

_____

_____

Magnetic Fields An invisible magnetic field surrounds a magnet. The magnetic field applies forces to other magnets or other magnetic material even when they are not in contact. Since magnetic fields are invisible, they must be detected by the forces they apply. In the photo on the right, iron filings are used to model a magnetic field around a bar magnet. The iron filings form a pattern of curved lines called magnetic field lines. The lines are closest at the magnet's poles, where the magnetic force is strongest. As the field lines become farther apart, the field and the force become weaker.

Compasses The needle of a compass is a small magnet. Like other magnets, a compass needle has a north pole and a south pole. If a compass needle is within any magnetic field, it will line up with the magnet's field lines. A compass needle does not point directly toward the poles of a magnet. Instead, the needle aligns with the field lines and points in the direction of the field lines.

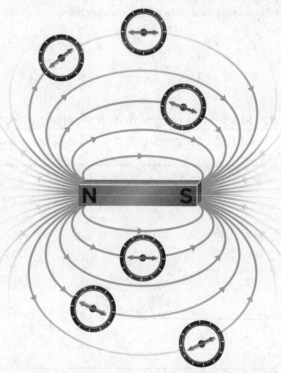

 THREE-DIMENSIONAL THINKING
Model a compass needle when it is near the north end of a magnet.

GO ONLINE for additional opportunities to explore!

Want to know more about magnetic fields? Investigate magnetic fields by performing one of the following activities.

☐ **Construct** a 3-D model of magnetic fields in the **Lab** *Experiment with Magnets.*

OR

☐ **Explore** how compasses work in the **PhET Interactive Simulation** *Magnet and Compass.*

Read a Scientific Text

EARTH SCIENCE › Connection A magnetic field surrounds Earth similar to the way a magnetic field surrounds a bar magnet. Earth's magnetic field is due to spinning molten iron and nickel in its outer core. Like all magnets, Earth has north and south magnetic poles. Earth's magnetic field protects Earth from cosmic rays and charged particles coming from the Sun. It pushes away some charged particles and traps others.

Earth's magnetic poles and geographic poles are not in the same spot. Read about how one scientist's job is to find the location of Earth's magnetic pole.

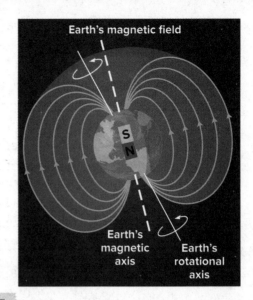

CLOSE READING

Inspect
Read the passage *Earth's Inconstant Magnetic Field.*

Find Evidence
Reread the passage. Underline the evidence that the Earth's magnetic poles change location over time.

Make Connections
Communicate With your partner, evaluate the evidence. What other evidence would you need to make a claim about how the Earth's magnetic poles change?

PRIMARY SOURCE

Earth's Inconstant Magnetic Field

Every few years, scientist Larry Newitt of the Geological Survey of Canada goes hunting. He grabs his gloves, parka, a fancy compass, hops on a plane and flies out over the Canadian arctic. Not much stirs among the scattered islands and sea ice, but Newitt's prey is there—always moving, shifting, elusive. His quarry is Earth's north magnetic pole.

At the moment it's located in northern Canada, about 600 km from the nearest town: Resolute Bay. … Newitt stops there for snacks and supplies—and refuge when the weather gets bad.

Scientists have long known that the magnetic pole moves. James Ross located the pole for the first time in 1831 after an exhausting arctic journey during which his ship got stuck in the ice for four years. … In 1904, Roald Amundsen found the pole again and discovered that it had moved—at least 50 km.

The pole kept going during the 20th century, north at an average speed of 10 km per year, lately accelerating "to 40 km per year," says Newitt. At this rate it will exit North America and reach Siberia in a few decades.

Keeping track of the north magnetic pole is Newitt's job. "We … check its location once every few years," he says. "We'll have to make more trips now that it is moving so quickly."

Earth's magnetic field is changing in other ways, too: Compass needles in Africa, for instance, are drifting about 1 degree per decade. And globally the magnetic field has weakened 10% since the 19th century. [...]

Sometimes the field completely flips. The north and the south poles swap places. Such reversals, recorded in the magnetism of ancient rocks, are unpredictable. They come at irregular intervals averaging about 300,000 years; the last one was 780,000 years ago. Are we overdue for another? No one knows.

Source: National Aeronautics and Space Administration

Magnetic Strength Magnets can vary in strength. For example, a magnet from your refrigerator can be used to pick up paper clips, but a much stronger magnet would be needed to lift a car. A strong magnet has a stronger magnetic field. This means that a magnetic object placed in the field will experience a stronger magnetic force. The strength of the magnetic field affects the use of the magnet. For example, the magnetic field used for a type of medical test called Magnetic Resonance Imaging (MRI) is about 200 times stronger than the magnetic field of a refrigerator magnet. This is why you must remove all magnetic objects before entering the MRI room.

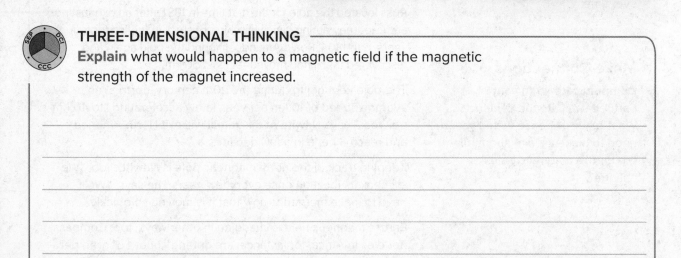

This MRI contains stronger magnets than the ones found in your kitchen.

THREE-DIMENSIONAL THINKING

Explain what would happen to a magnetic field if the magnetic strength of the magnet increased.

_____
_____
_____
_____
_____

COLLECT EVIDENCE

How do magnetic fields explain why magnets can keep a plant from falling? Record your evidence (B) in the chart at the beginning of the lesson.

What happens when magnetic fields interact?

Potential energy is energy stored in an object by reason of its position. In this activity, you will investigate how the potential energy of a magnet can be changed.

Do these magnets have energy?

LAB Moving Magnets

Safety

Materials

bar magnets (2)
tape
5-N spring scale

Procedure

1. Read and complete a lab safety form.

2. Place two bar magnets on a tabletop with the south poles facing one another. Push the two magnets toward one another until they are touching. Then release the magnets. Record your observations in the Data and Observations section on the next page.

3. Tape the hook of a spring scale securely to the south pole of one magnet. Place the magnet on the table with its north pole facing the south pole of the other magnet. Push the two magnets together until the ends are touching.

4. Holding the other magnet in place with one hand, slowly pull on the spring scale attached to the other magnet. Record the reading on the scale when the magnets move apart.

5. Hold the spring scale stationary and slowly push the south pole of the other magnet toward the north pole of the magnet with the spring scale. Observe and record the reading on the spring scale as the magnets move closer together.

6. Follow your teacher's instructions for proper cleanup.

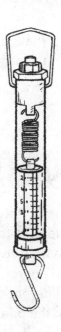

EXPLORE/EXPLAIN Lesson 1 Magnetic Forces

Data and Observations

Analyze and Conclude

7. Explain your observations in terms of potential and kinetic energy.

8. What effect does position have on the amount of potential energy stored in a magnet?

Magnetic Potential Energy To separate two opposite poles, energy has to be added to the magnetic field. Therefore, magnetic potential energy increases between the two attracting poles. **Magnetic potential energy** is stored energy due to the interactions of magnetic poles in a magnetic field. The difference in potential energy depends on the strength of the magnetic fields, the orientation of the poles, and the distance between the two poles.

As shown in the figure below, there are two ways to increase the magnetic potential energy between two magnets. When two similar poles are pushed together, the potential energy increases between the two magnets. When two opposite poles are pulled away from each other, the potential energy increases between the two magnets. In both cases, a force must be applied to increase the magnetic potential energy.

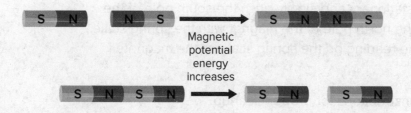

Transferring Magnetic Potential Energy When a magnetic field applies a strong enough force on an object to overcome the static forces on an object, the object moves. When a magnetic force causes an object to move, it does work on the object. When a paper clip moves toward a magnet, the magnetic field does work on the paper clip.

THREE-DIMENSIONAL THINKING

Think about when you moved two south poles together in the Lab *Moving Magnets*.

1. Sketch a **model** below that identifies the components of the **system**, the forces acting on the objects, the positions of the two objects, and the potential **energy** of the system.

2. **Explain** what happens to the amount of potential energy in the system above as the position of the magnets changes.

_____
_____
_____

3. What had to happen for the system above to gain energy?

_____
_____
_____

COLLECT EVIDENCE

How does magnetic potential energy explain why magnets can keep a plant from falling? Record your evidence (C) in the chart at the beginning of the lesson.

EXPLORE/EXPLAIN Lesson 1 Magnetic Forces 189

What makes a material act like a magnet?

Did you notice any of the paper clips sticking together after you removed them from the magnet in the Lab *Paper Clip Pick Up?* Sometimes, magnetic materials gain magnet-like properties. What causes materials to act like a magnet? Let's find out!

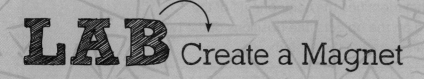

Safety

Materials

bar magnet   paper clip
nail   hard surface

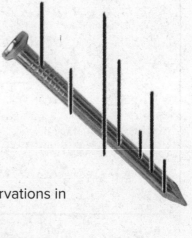

Procedure

1. Read and complete a lab safety form.

2. Touch each end of the nail to the paper clip. Record your observations in the Data and Observations section below.

3. Move the nail 25 times in the same direction across one end of the magnet. Repeat step 2, using the rubbed nail. Record your observations.

4. Drop the nail several times onto a hard surface. Repeat step 2 using the dropped nail. Record your observations.

5. Follow your teacher's instructions for proper cleanup.

Data and Observations

Analyze and Conclude

6. Were you able to get the nail to act like a magnet? Explain.

_____

_____

_____

_____

Magnetic Domains Remember, all matter is made of particles called atoms. Every atom has its own magnetic field. In some materials, atoms are grouped in magnetic domains. A **magnetic domain** is a region in a magnetic material in which the magnetic fields of the atoms all point in the same direction. The magnetic fields of the atoms within a domain combine into a single field around the domain.

Nonmagnetic Materials Most materials do not have atoms grouped in magnetic domains. Part A of the figure on the right shows how the magnetic fields of the atoms of a plastic comb point in many different directions. The magnetic effects of the random magnetic fields cancel each other. These nonmagnetic materials do not have any magnetic properties and cannot be made into magnets.

Magnetic Materials In magnetic materials, atoms are grouped in magnetic domains. However, not all magnetic materials are magnets. As shown in part B of the figure, the magnetic fields of the domains of the steel nail point in different directions. The magnetic fields of these domains cancel each other, so the magnetic material is not a magnet.

A magnetic material becomes a magnet as the magnetic fields of the material's magnetic domains line up in the same direction. Part C of the figure shows the aligned magnetic fields of the magnetic domains of a bar magnet. The magnetic fields of the domains combine to form a single magnetic field around the entire object. In this case, the magnetic material is a magnet.

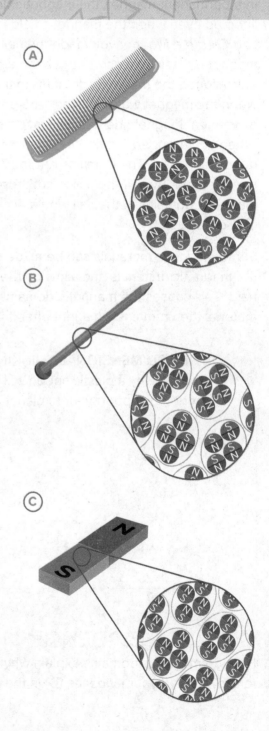

Temporary and Permanent Magnets Placing a magnetic material in a strong magnetic field causes the material's magnetic domains to line up. Examine the figure on the right. The nail is a magnet only when it is close to another magnet. There, the magnetic field is strong enough to cause the nail's magnetic domains to line up. However, when you move the nail away from the magnet, the domains in the nail will return to pointing in different directions. The nail will no longer attract other magnetic materials. This is a temporary magnet.

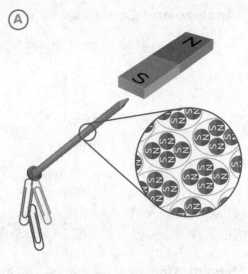

In a permanent magnet, the magnetic domains remain lined up even when the magnetic field is removed. In the Lab *Create a Magnet* you made the nail a permanent magnet by moving it across the bar magnet 25 times. This aligned the magnetic domains in the nail even when the magnet was removed. The domains in permanent magnets do not always remain aligned. When you dropped the nail, the magnetic domains fell out of alignment. The domains returned to a random arrangement. Even if the nail is not dropped, the magnetic domains in the nail would eventually shift out of alignment.

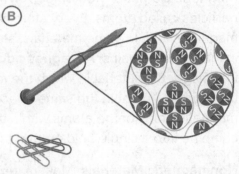

Some magnetic materials can be made into permanent magnets by heating the magnetic materials and allowing them to cool in a very strong magnetic field. This causes the magnetic domains to align and stay aligned. The material then remains a magnet after it is removed from the field.

THREE-DIMENSIONAL THINKING

Explain why a plastic spoon will never act like a magnet. Use a **model** to support your explanation.

COLLECT EVIDENCE

How do magnetic domains explain why magnets can keep a plant from falling? Record your evidence (D) in the chart at the beginning of the lesson.

A Closer Look: Magnetic Migration

LIFE SCIENCE Connection Each spring and fall, thousands of birds take wing, beginning their long flights to their seasonal nesting or feeding grounds. Some birds migrate the length of a continent! How do they make this trip season after season and end up at the same location?

Scientists are studying how birds use Earth's magnetic field to accurately traverse the continents. In one study, birds were moved from one coastal location where they were flying north, to a location hundreds of miles inland to the east. The birds were able to realign themselves northwest to the original heading. This indicates that birds have an internal compass that enables them to always know the direction they are heading.

To find out exactly how birds can detect Earth's magnetic field lines, scientists are taking a closer look at proteins found in migratory bird's eyes. They are hoping to find a clue on how birds can detect Earth's magnetic field.

It's Your Turn

Create a Blog Birds are not the only animals that migrate. How do salmon, monarch butterflies, deer, and even some bacteria use Earth's magnetic field to move? Research one of these or another question about Earth's magnetic field. Create a blog that explains what you found about Earth's magnetic field.

LESSON 1
Review

Summarize It!

1. **Organize** Create a graphic organizer that relates energy, magnetic fields, magnetic forces, magnetic and nonmagnetic materials, and permanent and temporary magnets.

Three-Dimensional Thinking

Examine the two magnetic field models below to answer questions 2–4.

X

Y

2. Which system best illustrates attractive forces?

   A X, because the magnetic field lines connect to each other

   B Y, because the magnetic field lines repel each other

   C X, because the magnetic field lines repel each other

   D Y, because the magnetic field lines connect to each other

3. Which system would need an external force to move the magnets closer together?

   A X, because the magnets are attracting

   B Y, because the magnets are repelling

   C X, because the magnets are repelling

   D Y, because the magnets are attracting

4. In which system could you increase the magnetic potential energy by pushing the two magnets closer together?

   A X, because the magnets are attracting

   B Y, because the magnets are repelling

   C X, because the magnets are repelling

   D Y, because the magnets are attracting

Real-World Connection

5. **Indicate** Plan an investigation to determine if a car door is made of a nonmagnetic material, such as plastic.

_____
_____
_____
_____

6. **Explain** You notice that a small magnet will stick to a refrigerator door, but a nail will not stick. What questions does this raise? Explain what causes the difference.

_____
_____
_____
_____
_____

 Still have questions?
Go online to check your understanding about magnetic fields and forces.

 **REVISIT PAGE KEELEY SCIENCE PROBES** Do you still agree with the statement you chose at the beginning of the lesson? Return to the Science Probe at the beginning of the lesson. Explain why you agree or disagree with that statement now.

START PLANNING
STEM Module Project Engineering Challenge

Now that you've learned about magnetic forces, go to your Module Project to determine the criteria and constraints that will guide your design. Keep in mind how you could incorporate magnets in the design of the sandbox crane for your friend.

 **EXPLAIN THE PHENOMENON** Revisit your claim on why magnets can keep a plant from falling. Review the evidence you collected. Explain how your evidence supports your claim.

196 EVALUATE Module: Electromagnetic Forces

LESSON 2 LAUNCH

Electric Charge

Electric charges surround you all the time. Some electric charges, such as lightning, are quite strong. Others are barely noticeable. What happens when two electrically charged objects get near each other? Circle the response that best matches your thinking.

A. The two objects will be pulled toward each other.
B. The two objects will be pushed away from each other.
C. The two objects will either be pulled together or pushed apart.
D. The charges will cancel out and neither object will move toward or away from the other.

Explain your thinking. Describe your ideas about electric charge.

_____
_____
_____
_____
_____
_____

You will revisit your response to the Science Probe at the end of the lesson.

LESSON 2
Electric Forces

198 ENGAGE Module: Electromagnetic Forces

ENCOUNTER THE PHENOMENON | Why does the balloon attract the water?

As a team, inflate a balloon and tie the end. Your partner will hold a funnel over a large bowl and pour a cup of water through the funnel. As the water gently flows, bring the balloon as close to the stream of water as you can without getting it wet. Record or illustrate your observations below. Next, rub the balloon on some fabric. Then, your partner will hold a funnel over a large bowl and pour a cup of water through the funnel again. As the water gently flows, bring the balloon as close to the stream of water as you can without getting it wet. Observe the interaction between the water and the balloon. Record and illustrate your observations below.

GO ONLINE
Watch the video *Move that water!* to see this phenomenon in action.

ENGAGE Lesson 2 Electric Forces 199

EXPLAIN
THE PHENOMENON

Have you ever pulled a sweater out of the clothes dryer and found other clothes clinging to it? Maybe you heard a crackling sound or even saw sparks when you pulled the items apart. Think about another time that you have noticed unseen attractive forces. Make a claim about what could be responsible for the attractive forces between the balloon and the water.

CLAIM
The water is attracted to the balloon because...

 COLLECT EVIDENCE as you work through the lesson. Then return to these pages to record your evidence.

EVIDENCE

A. What evidence have you discovered to explain the charges that charged particles can have?

B. What evidence have you discovered to explain how electric fields interact?

200 Module: Electromagnetic Forces

MORE EVIDENCE

C. What evidence have you discovered to explain how objects hold electric charges?

When you are finished with the lesson, review your evidence. If necessary, based on the evidence, revise your claim.

REVISED CLAIM

The water is attracted to the balloon because…

Finally, explain your reasoning for how and why your evidence supports your claim.

REASONING

The evidence I collected supports my claim because…

How do electric charges interact?

To open a door, your hand must touch the door to apply a force to it. However, a charged object does not have to touch another charged object to apply a force to it. What types of charges exist that can create these forces?

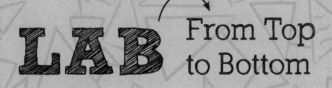

Safety

Materials

transparent tape   smooth surface

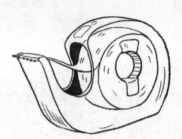

Procedure

1. Read and complete a lab safety form.

2. Fold over about 1 cm on the end of a roll of tape to make a handle.

3. Tear off a strip of tape about 10 cm long. Stick the strip to a smooth surface. Mark the handle with a *B*. This will be the bottom tape.

4. Make an identical piece of tape, and place it directly on top of the first tape strip. Mark the handle with a *T*. This will be the top tape.

5. Pull both pieces of tape off the surface together, and then pull them apart.

6. Move the non-sticky sides of both tapes close to each other. Record your observations on the next page.

7. Replace the bottom tape on the smooth surface, and place the top tape on top of the bottom tape. Pull both pieces of tape off the surface together, and then pull them apart.

8. With a partner, move the non-sticky sides of two top tapes close to each other. Then, move the non-sticky sides of two bottom tapes close to each other.

9. Record your observations on the next page.

202 EXPLORE/EXPLAIN Module: Electromagnetic Forces

10. Follow your teacher's instructions for proper cleanup.

Data and Observations

11. Fill in the table with *attract* or *repel* based on your observations.

	Top	Bottom
Top		
Bottom		

Analyze and Conclude

12. What rules for these attractions can you infer?

_____
_____

13. How are these results like magnetic forces? How are they different?

_____
_____
_____

14. Did this lab provide evidence to support the claim that forces can act on an object without touching the object? How could you improve the procedure to collect more data to support this claim?

_____
_____
_____
_____

Charges There are two types of electric charge—positive charge and negative charge. Oppositely charged particles attract each other. Similarly charged particles repel each other. How do charged objects apply electric forces to each other without touching? Scientists have found that there is a region around a charged object that applies an electric force to other charged objects. This invisible region around any charged object where an electric force is applied is an **electric field.** The electric force applied by an object's electric field will either attract or repel other charged objects. The electric force is a noncontact force. In the photo to the right, positive charges accumulate on the woman. The charges spread out and push away from each other. When the charges accumulate on hair, each hair will repel away from every other hair. This is why her hair sticks up.

Positively and negatively charged objects attract each other.

Two negatively charged objects repel each other.

Two positively charged objects repel each other.

THREE-DIMENSIONAL THINKING

When Jorge got to the bottom of the slide, his hair was sticking up. Make an **argument** on the **cause** and **effect** of accumulating charged particles that made his hair stand up.

_____
_____
_____
_____
_____

COLLECT EVIDENCE

How do the interactions of electric charges help explain why the balloon attracts the water? Record your evidence (A) in the chart at the beginning of the lesson.

 Want more information?
Go online to read more about electric forces from charged particles.

FOLDABLES
Go to the Foldables® library to make a Foldable® that will help you take notes while reading this lesson.

What determines the strength of an electric field?

An electric field surrounds charged particles. These fields are the reason that two objects do not need to touch for a force to act between them. Think back to the balloon. The charged particles on the balloon attract the water through an electric field. What determines how strong or weak the field is? Let's find out!

LAB Paper Pick Up

Safety

Materials

glass rod wool
newspaper scissors

Procedure

1. Read and complete a lab safety form.

2. Cut the paper into small pieces between 0.5 cm and 1 cm in size. Place the paper on a wooden tabletop or the cover of a book.

3. Hold the glass rod by one end and run it over the wool one or two times.

4. Quickly bring the end of the rod near the paper pieces without touching them. Observe what happens over the next 10–15 seconds, but do not touch the paper or the rod other than holding the end. When all movement has stopped, record your observations below.

EXPLORE/EXPLAIN Lesson 2 Electric Forces

Procedure, continued

5. Hold the rod by one end and run it through the wool ten times.

6. Quickly bring the end of the rod near the paper pieces without touching them. Observe what happens over the next 10–15 seconds. Record your observations.

7. Follow your teacher's instructions for proper clean up.

Analyze and Conclude

8. Explain what happened when you increased how many times you ran the glass rod through the wool.

9. Draw a series of diagrams of the rod and paper, using electric charges to explain what happened to cause each of the actions of the paper.
a) Draw what you think happened to the charges when you ran the rod through the wool.

b) Draw the charges when you brought the rod near the paper.

c) Draw the charges when the paper was touching the rod.

10. What evidence did you gain from changing the number of times you ran the rod through the wool?

_____
_____
_____

11. Why was changing the number of times you ran the glass rod through the wool necessary for measuring the electric force? How could you improve the procedure to collect more data to support the claim that increasing the number of times you ran the glass rod through the wool changes the electric force of the rod?

_____
_____
_____
_____

Electric Field Strength The strength of the electric force between two charged objects depends on two variables—the total amount of charge on both objects and the distance between the objects. When you run a comb through your hair, charged particles move from the comb to your hair. The negatively charged comb can attract paper just like the rod. The more charges on the comb, the stronger the force is on the paper. Distance also affects the strength of the electric force between electric charges. The larger the distance between the two objects, the weaker the force between them.

COLLECT EVIDENCE
How does the strength of the electric field help explain why the balloon attracts the water? Record your evidence (B) in the chart at the beginning of the lesson.

EXPLORE/EXPLAIN Lesson 2 Electric Forces

What happens when electric fields interact?

The strength of an electric field is dependent on the distance and the size of the charge. How would the field look if there were both positive and negative charges? How would their fields interact? Let's investigate!

INVESTIGATION

Field Rings

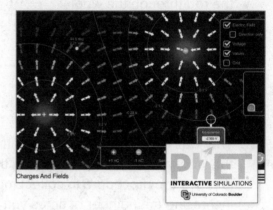

GO ONLINE Explore the PhET interactive simulation *Charges And Fields*.

After discovering the simulation on your own, refresh the settings and follow the instructions below.

1. Add 1 positively charged particle to the screen.
2. Use the equipotential sensor to find an area that has a reading of 2.0 V.
3. Click the pencil to draw a line that circles the charge.
4. Use the sensor to check everywhere on the line. What does this line represent?

_____

_____

5. Move the sensor toward the charged particle until the sensor has a reading of 2.5 V, and draw another line by clicking the pencil.
6. Move the sensor toward the charged particle until the sensor has a reading of 3.0 V, and draw another line by clicking the pencil.
7. Continue to draw lines every 0.5 V until the sensor has a reading of 10.0 V.
8. What relationship exists between distance and the strength of the sensor reading?

_____

_____

9. Refresh the settings and add 1 positively charged particle to the left side of the screen.

208 EXPLORE/EXPLAIN Module: Electromagnetic Forces

10. Add 1 negatively charged particle to the left side of the screen.

11. Use the equipotential sensor to find an area that has a reading of 2.0 V and click the pencil to draw a line that circles the charge.

12. Move the sensor to an area that has a reading of −2.0 V, and draw a line that circles the charge.

13. Move the sensor to an area that has a reading of 0.0 V, and draw a line.

14. Use the yellow sensor to test what the electric force would be at −2.0 V, 0.0 V, and 2.0 V. Record your observations below.

_____
_____
_____

15. Predict what would happen if a positive charge was placed between the two charges and allowed to move. Explain.

_____
_____
_____
_____

THREE-DIMENSIONAL THINKING

Ask questions about how the **data** gained from the PhET interactive simulation *Charges And Fields* would help you determine how the magnitude of the charge and the sign of the charge relate to the force between two charged objects. Record your questions in the space below.

_____
_____
_____
_____
_____

EXPLORE/EXPLAIN Lesson 2 Electric Forces

Electric Potential Energy The stored energy due to the interactions of charges in an electric field is electric potential energy. Consider two unlike charges. The charges attract each other. A force must be applied to keep the unlike charges apart. When this force is applied, energy is added to the electric charges and the electric potential energy between the two charges increases. The larger the charges or the greater the distance between the two charges, the greater the electric potential energy between the two charges. The figure on the right illustrates the similar nature of gravitational potential energy and electric potential energy.

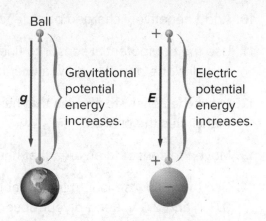

It is not the potential energy of the object at one particular point that is of interest. Often, scientists want to know the change in potential energy between two points.

Charged Objects Objects can gain or lose charged particles. Some materials hold their charged particles with less force than other materials. As a result, charged particles often move from one object to another. When this happens, the positive charge and negative charge on the objects become unbalanced. An unbalanced negative or positive electric charge on an object is sometimes referred to as a static charge. An **electrically charged** object has an unbalanced amount of positive charge or negative charge.

How do different materials hold electric charges?

The balloon at the beginning of the lesson held a charge that attracted the water. If you were to try the same experiment again with a piece of paper instead of the balloon, would the results be the same? How do you think different materials affect how the object holds a charge? Investigate with the online opportunities!

GO ONLINE for additional opportunities to explore!

Investigate how materials exchange and hold charges by performing one of the following activities.

- [ ] **Carry out** an investigation in the **Lab** *Guilty as Charged*.

  OR

- [ ] **Argue** how charges act on conductors and insulators after watching the **animation** *Transferring Electric Charge*.

Neutral Objects All objects contain charged particles. When an object has equal amounts of positive charge and negative charge, the charges are balanced. An object with equal amounts of positive charge and negative charge is **electrically neutral.** Electrically neutral objects do not attract or repel each other. Neutral objects, such as the paper from the Lab *Paper Pick Up,* can be attracted to charged objects. When the negatively charged rod came near the paper, the negative charges in the paper were pushed away. The positively charged particles in the paper became attracted to the rod.

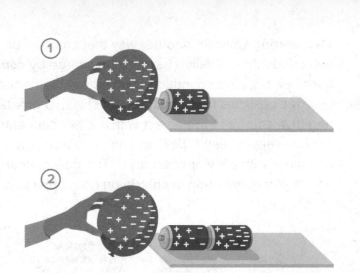
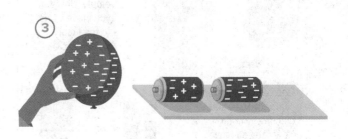

Look at the figure. When a negatively charged balloon is brought near a can, the negative charges in the can are pushed to the opposite end. When two cans are touching, negative charges are pushed to the farther can. Once the negative charges are in the farther can, the cans are separated. This leaves the negative charges in the farther can and the positive charges in the closer can. This method of charging an object without touching it is called **induction.**

Conductors and Insulators
To understand how charged particles can move from one object to another, you need to know about two basic types of materials—electric insulators and electric conductors. A material in which charges cannot easily move is an **electric insulator.** Glass, rubber, wood, and even air are good electric insulators. A material in which charged particles can easily move is an **electric conductor.** This is how a lightning rod works. The metal rod is a good conductor, so the charges can easily flow through the rod and safely to the ground. Most metals, such as copper and aluminum, are good electric conductors.

EXPLORE/EXPLAIN Lesson 2 Electric Forces 211

Transferring Charge Another way that charged particles transfer between two conductors is called transferring charge by **conduction**. As shown in the figure below, when conducting objects with unequal charges touch, charged particles flow from the object with a greater concentration of negative charge to the object with a lower concentration of negative charge. This is similar to water flowing from a container with a higher water level to a container with a lower water level. The flow of charged particles continues until the concentration of charge on both objects is equal.

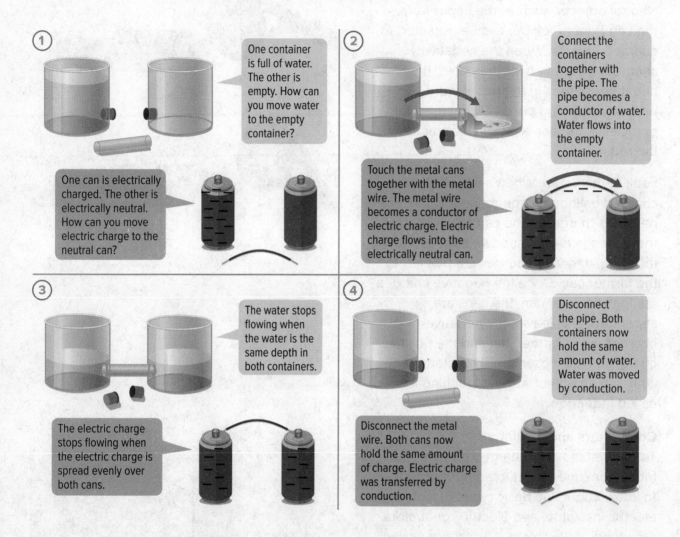

Conservation of Charge Notice that the amount of water did not change. The amount of water that started in the full container is the same as the amount of water after the two containers became equal. This is similar to the charged particles flowing between the two conductive objects. The total charge in a closed system does not change.

COLLECT EVIDENCE

How do the charges held on different materials help explain why the balloon attracts the water? Record your evidence (C) in the chart at the beginning of the lesson.

Van de Graaff Generator

HOW IT WORKS

How can a machine move negatively charged particles?

Have you ever seen a Van de Graaff generator? Originally, scientists studied atoms with these machines. Now, they are often seen in science museums, where they are used to demonstrate the effects of electric charge.

How does this device generate electric charge? Recall that some materials hold negatively charged particles more loosely than other materials. In a Van de Graaff generator, the metal dome holds negatively charged particles more loosely than the belt. As the belt travels over the top roller past the upper comb, negatively charged particles move from the metal dome through the upper comb and onto the belt. This leaves the dome positively charged.

The excess negatively charged particles on the belt move from the belt onto the lower comb as the belt moves over the roller at the bottom of the generator. The negatively charged particles then travel through a wire out of the machine and into the ground. This process of negatively charged particles moving from the dome to the ground continues as long as the generator is running.

Soon, the dome loses so many negatively charged particles that it acquires a very large positive charge. The positive charge on the dome becomes so great that negatively charged particles create an electric spark as they jump back onto the dome from any object that will release them. That object could be you if you stand close enough.

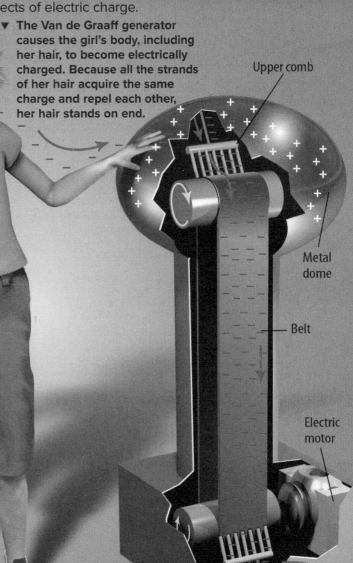

The Van de Graaff generator causes the girl's body, including her hair, to become electrically charged. Because all the strands of her hair acquire the same charge and repel each other, her hair stands on end.

Upper comb
Metal dome
Belt
Electric motor
Lower comb
Ground

It's Your Turn

Report Find out how to build a simple Van de Graaff generator using everyday materials. Share your findings with your classmates in a slide show presentation.

ELABORATE Lesson 2 Electric Forces

LESSON 2
Review

Summarize It!

1. **Organize** Make a model that relates the sign and magnitude of a charge to the attractive forces between objects.

Three-Dimensional Thinking

While doing laundry, Jaimee pulls clothes from the dryer. Some socks cling to a blanket.

2. Which statement explains why the socks cling to the blanket?

   A The socks and blanket dried together which caused them to cling to each other.

   B The socks and blanket are conductors that picked up some positive charges that keep the clothes together.

   C The socks and blanket are insulators that picked up some negative charges that keep the clothes together.

   D The clothes picked up opposite charges. The opposite charges are attracted to each other.

3. Which solution would reduce this problem in a dryer?

   A A crinkled-up ball of aluminum foil will conduct the charges and remove the charges from the clothes.

   B Place less clothes in the dryer to prevent rubbing.

   C Lower the electricity the dryer uses to lower the number of charges in the dryer.

   D Dry clothes without them touching so that the clothes do not dry together.

Real-World Connection

4. **Infer** A ceiling fan's blades collect dust faster than other objects in a room. The constant pushing and rubbing against the air provides a cool breeze for the people in the room. What causes the ceiling fan's blades to collect dust so quickly?

___

5. **Explain** Electric cables for your lamp have a plastic or rubber coating around the wire. Explain why this material surrounds the metal wires.

___

 Still have questions?
Go online to check your understanding about electric forces from charged particles.

REVISIT PAGE KEELEY SCIENCE PROBES

Do you still agree with the statement you chose at the beginning of the lesson? Return to the Science Probe at the beginning of the lesson. Explain why you agree or disagree with that statement now.

EXPLAIN THE PHENOMENON

Revisit your claim about why the water is attracted to the balloon. Review the evidence you collected. Explain how your evidence supports your claim.

KEEP PLANNING

STEM Module Project Engineering Challenge
Now that you've learned about charged particles and electric forces, go back to your Module Project to construct an argument on which forces might be present in a metal pickup machine. Keep in mind how the balloon attracts the water when planning your device.

LESSON 3 LAUNCH

Plugging In

Amory turns the lights on in her house. Looking at the lights, she begins to wonder how the charges are moving through the bulb. Which statement best describes how the charges move through the bulb?

A. The bulb uses up the charges in producing light.
B. Charges clash in the bulb and produce light.
C. When there are charges in the bulb, light is produced.
D. When charges pass through the bulb, light is produced.

Which statement do you most agree with? Explain your reasoning.

_____
_____
_____
_____
_____
_____

You will revisit your response to the Science Probe at the end of the lesson.

LESSON 3
Simple Circuits

ENCOUNTER THE PHENOMENON | How does a circuit make the light bulb light?

Your teacher will provide you with some circuits. In the space below, diagram each attempt, identifying if the light bulb will light or not. Explain why you think the circuits worked or not.

GO ONLINE
Check out *Simple Circuits* to see this phenomenon in action.

ENGAGE Lesson 3 Simple Circuits

EXPLAIN
THE PHENOMENON

Many common and useful circuits include only a few components. Simple circuits are found in flashlights, doorbells, and many kitchen appliances. Think about what makes these circuits work. Make a claim about how circuits can make the light bulb light up a dark room.

CLAIM
A light bulb lights because a circuit...

 COLLECT EVIDENCE as you work through the lesson. Then return to these pages to record your evidence.

EVIDENCE
A. What evidence have you discovered to explain how charged particles flow through a circuit?

Module: Electromagnetic Forces

MORE EVIDENCE

B. What evidence have you discovered to explain what factors affect an electric current?

When you are finished with the lesson, review your evidence. If necessary, based on the evidence, revise your claim.

REVISED CLAIM

A light bulb lights because a circuit...

Finally, explain your reasoning for how and why your evidence supports your claim.

REASONING

The evidence I collected supports my claim because...

Lesson 3 Simple Circuits

How do electric particles flow?

The light bulb requires a connection to a source of charged particles. Modern connections can be very complicated, but all connections are based on simple circuits. What connections do these circuits need to make a light bulb light?

LAB Lighten Up

Safety

Materials

D-cell battery   paper clips (4)

small light bulbs (2)

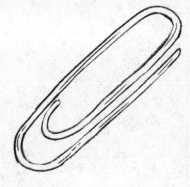

Procedure

1. Read and complete a lab safety form.

2. Examine a D-cell battery. Notice the difference between the two ends. Record your observations in the Data and Observations section below.

3. Using the paper clips, one small light bulb, and the battery, design a way to light the bulb. Record your observations below.

4. Find a way to light two bulbs using the paper clips and battery. Record your observations below.

5. Follow your teacher's instructions for proper cleanup.

Data and Observations

One light bulb	Two light bulbs

6. Draw a model of the circuit you made in step 3.

7. Draw a model of the circuit you made in step 4.

Analyze and Conclude

8. What patterns do you notice about what is required to light the light bulb?

_____

_____

_____

9. Why do you think the battery is required to light the light bulb?

_____

_____

_____

EXPLORE/EXPLAIN Lesson 3 Simple Circuits

THREE-DIMENSIONAL THINKING

Thorin noticed that when the paper clips came apart, the *effect was that the* light bulb went out. **Design a solution** for a circuit that must be able to turn off and on while connected to an electric energy source.

_____
_____
_____
_____

Simple Circuits All simple circuits contain a source of electric energy, an electric device, and an electric conductor. The source of electric energy in the Lab *Lighten Up* was the battery. The electric device was the light bulb, and the electric conductor was the paper clips. In addition to these basic components of all circuits, a switch is often included in a circuit. When a circuit is complete and electric energy flows through the circuit, it is called a **closed circuit.** An **open circuit** is a circuit that is not complete and no electric energy flows through the circuit. A switch changes a circuit between open and closed.

Charged Particles A D-cell battery contains chemicals that react and let off charged particles. Recall that charged particles repel like charges. The positive side of the battery has positively charged particles. Likewise, the negative side of the battery contains negatively charged particles. These charges travel along the conducting wire by the repulsive force between the like charged particles and the attraction between unlike charges. When the electrical charges pass through the electric device, electrical energy is transferred. The movement of electrically charged particles is an **electric current.**

Want more information?
Go online to read more about simple circuits.

FOLDABLES
Go to the Foldables® library to make a Foldable® that will help you take notes while reading this lesson.

COLLECT EVIDENCE

How does the flow of charges in a circuit help explain how a light bulb lights? Record your evidence (A) in the chart at the beginning of the lesson.

What factors affect an electric current?

A current flows through the conductive material that lights the bulb. The bulb is the electric device that converts the electric energy into light. The more energy, or current, the more light will be produced. What affects how much current there is in a wire?

OPEN INQUIRY LAB Power Up

Safety

Materials

D-cell batteries (2)   battery holders (2)
3-V flashlight bulb   bulb base
alligator clip wires (4)

Procedure

1. Read and complete a lab safety form.

2. Examine a D-cell battery. Determine the difference between the two ends and record the stated voltage. Record your observations.

_____

_____

3. Write a set of procedures in your Science Notebook that you will use to determine the relationship between voltage and current. Include in your procedure:

    A. The purpose of your investigation. Identify what relationship between the variables will be tested.

    B. The voltages to be used in the investigation.

    C. How current will be measured.

    D. The evidence needed to relate voltage and current.

    E. Needed materials and safety precautions to be taken.

EXPLORE/EXPLAIN Lesson 3 Simple Circuits

Procedure, continued

4. Use the table below to record your predictions for each circuit.

Circuit Description	Sketch of Circuit	Predicted Bulb Brightness	Observed Brightness
one battery and one bulb			
two batteries and one bulb version #1			
two batteries and one bulb version #2			

5. Construct and observe the circuits described in the table.

6. Compare the predicted brightness of the bulb in each circuit to the observed brightness.

7. Follow your teacher's instructions for proper cleanup.

Analyze and Conclude

8. What relationship exists between the voltage and current?

_____

_____

_____

9. What would happen to the brightness if you connected a battery with a stated voltage of 3 V?

_____

_____

226 EXPLORE/EXPLAIN Module: Electromagnetic Forces

Voltage You probably have heard the term *volt*. You use 1.5 V batteries in a flashlight. You plug a hair dryer into a 120 V outlet. But, what does this mean? In the Lab *Power Up*, a battery created an electric current in a closed circuit. Energy stored in the battery moves charged particles in the circuit. As the charged particles move through the circuit, the amount of energy transformed by the circuit depends on the battery's voltage. Recall that electrical potential energy is stored energy due to the interactions of charges in an electric field. **Voltage** is the electrical potential energy difference between two places on a circuit.

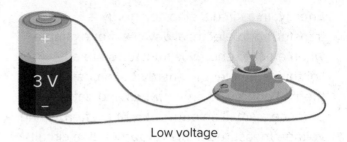
Low voltage

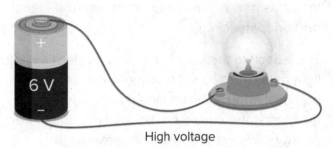
High voltage

The measure of the potential energy difference is made in volts (V). For example, a 9 V battery has a potential energy difference of 9 V between the positive and negative side. A circuit with a high voltage source transforms more electric energy to other energy forms than a circuit with a low voltage source. For example, a light bulb connected to a 9 V battery produces about six times more light and thermal energy than the same light bulb connected to a 1.5 V battery.

THREE-DIMENSIONAL THINKING

Isabella uses a voltmeter to measure the voltage of a battery. The voltage reads 9 V when she measures both sides. Make an **argument** about what the reading will be if she measures the same side of a battery. Use evidence to support your argument.

_____

_____

_____

_____

_____

_____

_____

_____

EXPLORE/EXPLAIN Lesson 3 Simple Circuits

Energy In a circuit, electric energy is transferred to electrical devices. For example, when electric energy is transferred to a lightbulb, the electric energy becomes light and thermal energy. Even the wires and batteries produce a small amount of thermal energy. The voltage measured across a portion of a circuit indicates how much energy is transferred in that portion of the circuit.

Think back to the light bulb. Most modern light bulbs are 120 V. This means that the potential energy will change by 120 V when measuring the circuit around the light bulb. The energy transfers from electrical energy to thermal energy and light. Power companies responsible for entire cities or more require much higher voltages to meet energy needs. The high voltage power lines that come from a power station are over 110,000 volts!

You have learned that electric conductors are materials, such as copper and aluminum, in which charged particles easily move. Usually electric wires are made of copper because copper is one of the best conductors.

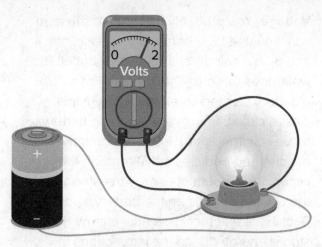

Higher voltage across light bulb

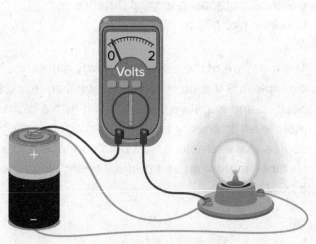

Lower voltage across wire

GO ONLINE for additional opportunities to explore!

Want to know more about how the material or electric device affects the current in a circuit? Investigate how changing materials and electrical devices changes the current by performing one of the following activities.

☐ **Carry out** an investigation in the **Lab** *One by One*.

OR

☐ **Use** mathematical thinking in the **PhET Interactive Simulation** *Ohm's Law*.

COLLECT EVIDENCE

How do the factors that affect electric current help explain how a circuit lights a light bulb? Record your evidence (B) in the chart at the beginning of the lesson.

SCIENCE & SOCIETY

A Smart Grid?

Electric Energy for the Future

The North American power grid is a system of interconnected electric transmission wires that reaches across the continent. This network of transmission wires includes smaller, regional grids in the eastern United States, the western United States, and Texas. The grid is the electric super highway that delivers electric energy to all our communities.

The grid, shown as red lines on the map, is aging quickly. Many transmission wires are too small to carry all the electric energy people demand. Parts of the grid often are overloaded. As electric current becomes too great in one part of the grid, that part shuts down to prevent damage to generators and transmission wires. The electric current then shifts to other transmission wires that become overloaded, too. This type of cascading overload can cause power failures and blackouts over large areas of the country. One solution to our electric distribution problem is to build a smart grid, shown as green lines on the map.

Computers at distribution centers throughout the grid would constantly analyze electric energy needs across the country. The system could route electric power from where it is produced, anywhere in the country, to where it is needed.

Also, consumers would have smart meters at their homes. Each smart meter would be connected to a personal computer to allow homeowners to see how much energy each of their household's electrical devices use. They quickly could see where they unwisely use electric energy. People could adjust their use of electric energy to save money and decrease their demands on the grid.

It's Your Turn

Research and Report Many experts agree that we must soon build a smart grid for electric energy distribution. Research why a smart grid is necessary for the development of alternative energy sources and create a pamphlet based on your findings.

LESSON 3
Review

Summarize It!

1. **Diagram** Make a model of an electric circuit that includes an electric energy source, an electric device, and a conductor.

Three-Dimensional Thinking

Use the model below to answer questions 2–3.

A 1.5 V battery is connected to a light bulb with some wires. One of the wires is cut, breaking the circuit.

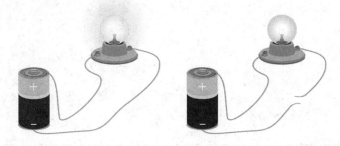

2. What is the electrical potential energy difference across the light bulb after the wire is cut?

   A 0.0 V

   B 1.5 V

   C −1.5 V

   D Need more information.

3. After the wire is cut, what is the electrical potential energy difference across the two ends of the cut wire?

   A 0.0 V

   B 1.5 V

   C −1.5 V

   D Need more information.

Real-World Connection

4. **Explain** A D-cell battery keeps a flashlight on all night. In the morning, you notice that the flashlight has gone out. Plan an investigation to determine the potential difference across the battery. Explain what you would expect to find and why.

_____

_____

_____

_____

5. **Explain** All voltmeters have two probes attached to make a measurement. Explain why you cannot make a voltmeter with a single probe to measure the voltage of a wire.

_____

_____

_____

_____

 Still have questions?
Go online to check your understanding about simple circuits.

 **REVISIT SCIENCE PROBES** Do you still agree with the statement you chose at the beginning of the lesson? Return to the Science Probe at the beginning of the lesson. Explain why you agree or disagree with that statement now.

 **EXPLAIN THE PHENOMENON** Revisit your claim about how a circuit lights a light bulb. Review the evidence you collected. Explain how your evidence supports your claim.

KEEP PLANNING
STEM Module Project Engineering Challenge

Now that you've learned about simple circuits, go back to your Module Project to model possible ways a circuit could be incorporated into the sandbox crane. Keep in mind how simple circuits light a light bulb.

232 EVALUATE Module: Electromagnetic Forces

LESSON 4 LAUNCH

Charged Magnets

Tony noticed that his speaker was magnetic. He wondered if electricity and magnetism were related. Circle the answer that best matches your thinking.

- **A.** There is no relationship between electricity and magnetism.
- **B.** There is a relationship between electricity and magnetism.
- **C.** There is a relationship between electricity and magnetism, but only when two objects are touching.

Explain your thinking.

_____
_____
_____
_____
_____
_____
_____

You will revisit your response to the Science Probe at the end of the lesson.

LESSON 4
Electromagnetism

234 ENGAGE Module: Electromagnetic Forces

ENCOUNTER THE PHENOMENON

Why does a metal detector work when the current is on but not when the current is off?

GO ONLINE

Watch the video *Jumping Magnets* to see this phenomenon in action.

Have you ever seen a metal detector being used? When a metal detector is turned on and is moved over a metal object, the detector will beep. How do you think a metal detector works? Watch the video *Jumping Magnets* and record your observations. What happens when current flows through the wire in the video? What do you think happens when current flows through the wires in a metal detector? With a partner, discuss ways in which the current may affect objects around it. Next, develop a list of questions that use scientific practices to help you find out more about what is happening.

ENGAGE Lesson 4 Electromagnetism **235**

EXPLAIN THE PHENOMENON

What do a metal detector and the wire from the video *Jumping Magnets* have in common? Both show that electricity and magnetism are connected. But how are they connected? Using your understanding of electric currents and magnetic forces, make a claim about why the metal detector works when the current is on but not when the current is turned off.

CLAIM
The metal detector works when the current is on but not when the current is turned off because…

 COLLECT EVIDENCE as you work through the lesson. Then return to these pages to record your evidence.

EVIDENCE
A. What evidence have you discovered to explain the existence of magnetic fields around a current-carrying wire?

B. What evidence have you discovered to explain how electric motors are used to produce motion?

236 Module: Electromagnetic Forces

MORE EVIDENCE

C. What evidence have you discovered to explain how a magnet can produce a current in a wire?

D. What evidence have you discovered to explain how electric generators produce electrical energy?

When you are finished with the lesson, review your evidence. If necessary, based on the evidence, revise your claim.

REVISED CLAIM
The metal detector works when the current is on but not when the current is turned off because…

Finally, explain your reasoning for how and why your evidence supports your claim.

REASONING
The evidence I collected supports my claim because…

Lesson 4 Electromagnetism

How do currents create magnetic fields?

Recall that a current is a stream of charged particles, and that a magnetic field surrounds a magnet. In the video *Jumping Magnets*, the wire hung between the poles of a magnet. The wire moved when the electricity was turned on. Is there a relationship between magnets and electricity? Let's investigate!

> **Want more information?**
> Go online to read more about how electric and magnetic forces are related.

LAB Pointing Directions

Safety

Materials

clamp attachment
ring stand
push pin
tape
alligator clip wires (4)
D-cell battery holders (2)
copper rod
ring
sharp pencil
small compasses (4)
D-cell batteries (2)
10-cm square piece of cardboard

Procedure

1. Read and complete a lab safety form.

2. With a clamp attachment, hang the copper rod through the center of a ring on a ring stand.

3. Use the push pin and the pencil to make a small hole in the center of the piece of cardboard. Slide the rod through the hole. Rest the cardboard on the ring. Secure the cardboard to the ring with tape.

4. Place four small compasses on the cardboard in a circle around the rod as shown in the figure to the right.

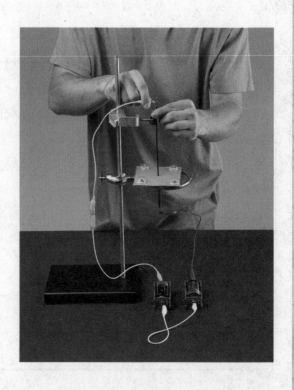

238 EXPLORE/EXPLAIN Module: Electromagnetic Forces

5. In the Data and Observations section below, draw your setup. Include the rod through the cardboard and the circle of compasses in your drawing. Indicate the direction that each compass points.

6. Use alligator clip wires to connect two D-cell batteries in holders in a series with the ends of the rod. Draw a second diagram. Show the direction the compasses point with the batteries connected.

   CAUTION: *Unhook the alligator clips from the rod after a few seconds to prevent it from overheating!*

7. Follow your teacher's instructions for proper cleanup.

Data and Observations

Set Up Before Attaching Batteries	Set Up After Attaching Batteries

Analyze and Conclude

8. Why do you think the compass needles behaved as they did after the batteries were connected?

_____

_____

EXPLORE/EXPLAIN Lesson 4 Electromagnetism

Moving Charges and Magnetic Fields A magnetic field surrounds an electric current. This is why a compass needle moves when placed near a current-carrying wire. The needle moves because the magnetic field around the wire applies a force to the compass needle. The magnetic field can also be seen with iron filings as shown in the photo.

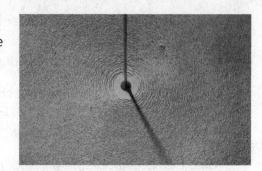

A magnetic field actually surrounds all moving charged particles. The magnetic fields of the flowing charges in a current-carrying wire combine to produce a magnetic field around the wire. If the current is turned off, the charged particles in the wire stop moving. Because only moving charged particles have a magnetic field, the magnetic field goes away when the particles stop moving.

The magnetic field around a wire is perpendicular to the current. This is seen as concentric circles around the wire. If a compass were held parallel to the wire, it would not respond to the magnetic field produced by the wire. The magnetic field is strongest closer to the wire. The farther away from the wire, the weaker the magnetic field becomes.

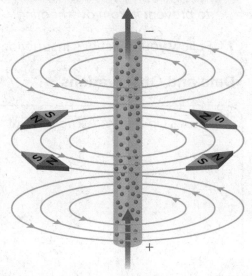

What affects the strength of a magnetic field around a current?

Are all magnetic forces around electric charges the same? You know you can increase a current by increasing the voltage across the wire. Does this affect the magnetic field the wire produces? Are there other ways to increase the magnetic field surrounding a current-carrying wire? Let's find out!

INVESTIGATION

Making Magnetic Fields

How can you increase the magnetic field around a wire?

GO ONLINE Explore the PhET interactive simulation *Magnets and Electromagnets*.

After exploring the simulation on your own, reset the simulation and follow the instructions on the next page.

240 EXPLORE/EXPLAIN Module: Electromagnetic Forces

1. Click on the Electromagnet tab. Click "Reset All".

2. Check the "Show Field Meter". All the boxes should now be checked. Look at the Field Meter. Move the Field Meter around the screen. The reading $\bar{B}$ illustrates the magnetic field strength. Complete the graphic organizer to describe the strength of the magnetic field near the wire coils.

| The closer to the wire coils | the _____ the magnetic field. |

3. Determine the cause-and-effect relationship between the magnetic field strength and the circuit. Move the Field Meter on top of the wire coils. Try to decrease the magnetic field strength without moving the circuit or the Magnetic Field Meter. Describe what you did to decrease the magnetic field strength.

_____

_____

_____

Increasing Magnetic Field Strength Around a Wire Magnetic fields around current-carrying wires are strongest closest to the wire. As you move away from the wire, the magnetic field becomes weaker.

The field around the wire becomes stronger as the current in the wire increases, or as more charged particles flow in the wire. Recall you can increase the current by increasing the voltage across the wire.

If a current-carrying wire is wound into a coil, the magnetic field becomes stronger. The magnetic fields around the individual loops of the coil combine, making the magnetic field stronger. Examine the figure on the right. The magnetic field lines are still perpendicular to the wires. They now form loops around the coils. The more coils in the wire, the stronger the magnetic field. This same principle applies if there are a larger number of straight wires all running parallel. The more wires running through a certain area increases the magnetic field surrounding the wires.

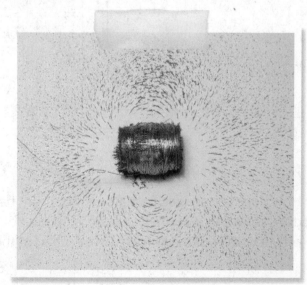

EXPLORE/EXPLAIN Lesson 4 Electromagnetism

Electromagnets You discovered in the Investigation *Making Magnetic Fields* that when you increase the number of coils in a current-carrying wire, the magnetic field around the coil increases. Another way to increase the strength of a magnetic field occurs when a magnetic material is placed within the coil. A temporary magnet made with a current-carrying wire coil wrapped around a magnetic core is called an **electromagnet.**

An electric current in a wire produces a magnetic field around the wire.

An electric current in a wire coil produces a magnetic field with a north pole and a south pole.

Placing an iron core within the coil greatly intensifies the magnetic field. This device is an electromagnet.

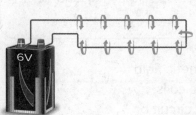
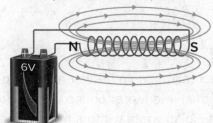
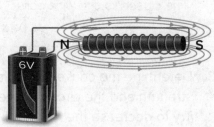

THREE-DIMENSIONAL THINKING

Model three **cause-and-effect** relationships that affect the strength of magnetic forces around current-carrying wires.

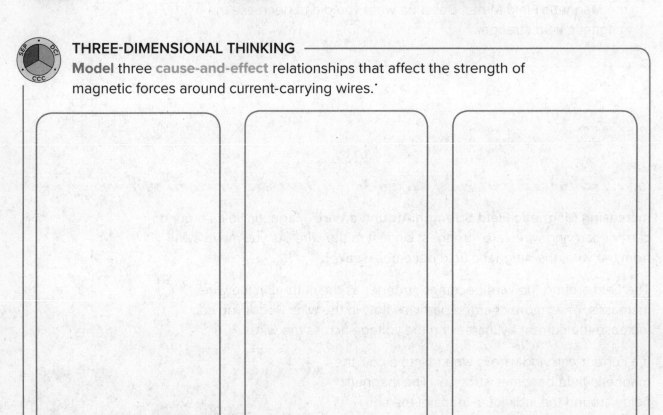

COLLECT EVIDENCE

How does a magnetic field around a current help explain why the metal detector works when the current is on but not when the current is turned off? Record your evidence (A) in the chart at the beginning of the lesson.

242 **EXPLORE/EXPLAIN** Module: Electromagnetic Forces

What makes electromagnets useful?

Electromagnets can be found in many devices, such as speakers, doorbells, brakes in cars, and large lifting cranes. To make an electromagnet you need a closed circuit with an electricity source and a wire coil with a magnetic material inside the coil. That seems like a lot of work just to make a magnetic field. You can get a magnetic field just from a magnet. Why would anyone use an electromagnet? What makes electromagnets so useful? Try your hand at making one to find out!

FOLDABLES
Go to the Foldables® library to make a Foldable® that will help you take notes while reading this lesson.

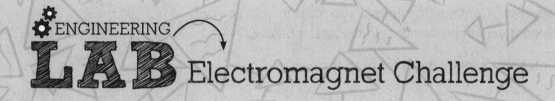

ENGINEERING LAB: Electromagnet Challenge

Safety

Materials

sandpaper
paper clips (20)
alligator clip wires (4)
iron nails (5)
bolts (5)
metric ruler
wire (150 cm)

cardboard tube
bar magnet
D-cell batteries (3)
D-cell battery holders (3)
balance
hot mitts

Procedure

1. Read and complete a lab safety form.

2. With sandpaper, rub off 2 cm of insulation from both ends of 150 cm of wire.

3. Make a coil by wrapping half of the wire around a nail. Leave a 5-cm tail of wire as you begin wrapping the wire. Approximately 75 cm of wire will remain at the other end of the coil.

4. Count how many paper clips the coil picks up and how it interacts with a permanent magnet. Record your observations in the Data and Observations section on the next page.

5. Use alligator clip wires to connect the tails of the coil to a D-cell battery in a holder. Repeat step 4. Record your observations.

   CAUTION: *Unhook the wire after a few seconds to avoid overheating!*

EXPLORE/EXPLAIN Lesson 4 Electromagnetism **243**

Procedure, continued

6. Measure and record the mass of the electromagnet assembly (not including the battery). Repeat step 4. Measure and record the mass of the paper clips you were able to pick up.

7. With your group, discuss how you can make your electromagnet stronger while keeping its mass as low as possible. Your goal is to build an electromagnet that will pick up the greatest mass of paper clips. The greater the mass of the paper clips (compared to the mass of the electromagnet), the more successful you are in your design.

8. When you have decided on your design, build and test it with your teacher's approval. Continue to improve your design until you are satisfied. Record your improvements in the space below.

9. Follow your teacher's instructions for proper cleanup.

Data and Observations

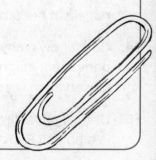

Analyze and Conclude

10. Why did the challenge have you change the strength of the magnetic field?

11. What factors did you change to increase the strength of the electromagnet?

12. Which factors were the most effective in increasing the strength of the electromagnet without greatly increasing its mass?

13. **MATH Connection** What was the ratio of the mass of paper clips you picked up to the mass of the electromagnet?

14. What are some benefits to using an electromagnet versus a permanent magnet?

Electromagnets Electromagnets are useful because they can be controlled in ways other magnets cannot. First, an electromagnet's magnetic field can be turned off and on. Turning off the electric current in the coil turns off the magnetic field. Examine the strong electromagnet in the figure on the right. This electromagnet is used to separate magnetic materials from nonmagnetic materials. If it were permanent, it could not let go of the magnetic materials. Once it is turned off, the electromagnet will lose its magnetic field so it can release the magnetic materials.

This electromagnet can be turned on and off.

Second, the strength of an electromagnet can be controlled by the number of loops in the coil and the amount of electric current in the coil. And finally, the north and south poles of the electromagnet reverse if the current reverses. You will find out how this is useful in the next section.

THREE-DIMENSIONAL THINKING

Three electromagnetic devices were designed. Analyze each device. Then, using your understanding of the cause-and-effect relationships that affect the strength of magnetic forces around current-carrying wires, circle which device would pick up the most paper clips. Explain your choice.

Device 1	Device 2	Device 3
20 coils of wire (no core) connected to two batteries	20 coils of wire with a nail in the core, connected to two batteries	20 coils of wire with a wooden core, connected to two batteries

How can electric energy be used to create motion?

You might recall that the north and south poles of an electromagnet reverse if the current reverses. This is the principle behind producing motion from electrical energy. Try your hand at producing motion using an electromagnet in the next lab.

246 EXPLORE/EXPLAIN Module: Electromagnetic Forces

LAB Motor On

Safety

Materials

sandpaper
large paper clips (2)
alligator clip wires (2)
push pins (4)
metric ruler

wire (150 cm)
15-cm X 15-cm foam board
magnet
D-cell battery
D-cell battery holder

Procedure

1. Read and complete a lab safety form.

2. Wrap the wire around a D-cell battery. Leave 5-cm tails at each end of the coil.

3. Wrap the tails once around the coil so that the coil is held together and the wires stick straight out perpendicular to the coil. See the top image on the right as an example. Hold the tails and spin the coil between your fingers. It should spin easily and not feel lopsided.

4. Lay the coil flat on the table. Using sandpaper, scrape the insulation off the visible side of the tails. Flip the coil over and scrape the insulation off one of the two tails. See the top image on the right as an example. Do not scrape the insulation off the coil!

5. Unfold each paper clip to form two S shapes. Using push pins, attach the paper clips to the board, as shown in the bottom image on the right.

6. Place the magnet on the board, between the paper clips.

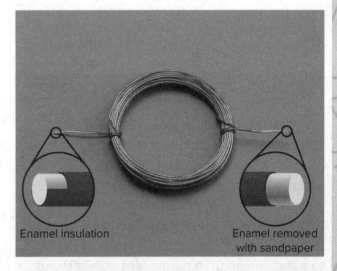

Enamel insulation — Enamel removed with sandpaper

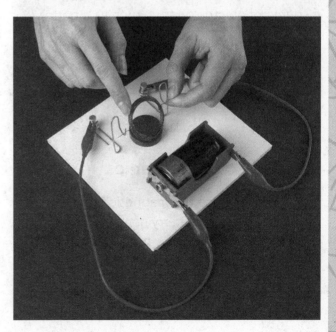

EXPLORE/EXPLAIN Lesson 4 Electromagnetism

Procedure, continued

7. Suspend the coil by its tails in the hooks formed in the paper clips.

8. Using alligator clip wires, connect the paper clips to the terminals of the D-cell's holder. Give the coil a twist, and watch it spin rapidly. If your motor does not spin, be sure that the insulation has been scraped correctly and that the bare metal touches the paper clips.

9. Follow your teacher's instructions for proper cleanup.

Analyze and Conclude

10. How does placing the coil of wires over the permanent magnet keep the coil spinning?

___

Magnets and Electric Motors Power tools, electric fans, hair dryers, computers, and even microwave ovens use electric motors. An **electric motor** is a device that uses an electric current to produce motion. A simple electric motor has three main parts. The main parts of an electric motor are a coil of wire connected to a rotating shaft, a permanent magnet, and a source of electric energy, such as a battery.

Some electric motors require a commutator. A commutator is a type of electrical switch that reverses the current in the coil.

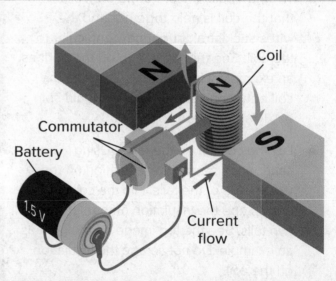

INVESTIGATION

Electric Motor Mechanics

Electric motors use the magnetic force properties of current-carrying wires to produce motion. Let's find out how they work.

GO ONLINE Watch the animation *Electric Motors*. After you have watched the animation, complete the graphic organizer on the next page.

248   EXPLORE/EXPLAIN Module: Electromagnetic Forces

Cause	Effect
Current supplied to the coil transforms the coil into an electromagnet.	
Coil is positioned between poles of a permanent magnet.	
The coil is on a rotating shaft.	
The direction of the current is reversed.	
Like poles of the magnets are aligned.	
The commutator continuously reverses the current.	
	The motor spins.

Using Electric Motors In an electric motor, electric energy is transformed to mechanical energy to produce motion. Electric motors are used in many devices from windshield wipers to CD players and even power windows on cars. Almost any device that needs to produce motion uses an electric motor. Electric motors can be tiny or very large. The strength of an electric motor depends on the strength of the permanent magnet, the voltage, and the number of wire coils.

COLLECT EVIDENCE

How do electric motors produce motion? Record your evidence (B) in the chart at the beginning of the lesson.

How can magnets produce an electric current?

You have seen that an electric current will produce a magnetic field. What if a magnet were moved near a wire without a power source. Would anything happen? Let find out!

LAB Coiled Up

Safety

Materials

cardboard tube
wire (300 cm)
ammeter
bar magnet
sandpaper
metric ruler

Procedure

1. Read and complete a lab safety form.

2. Wrap the wire around the cardboard tube to make a coil of about 20 turns. Remove the tube from the coil.

3. Use the sandpaper to remove 2 cm of insulation from each end of the wire.

4. Connect the ends of the wire to an ammeter. An ammeter measures the current in the wire. Record the reading from your meter in the Data and Observations section on the next page.

5. Insert one end of the magnet into the coil and then remove it. Record the current measured by the meter.

6. Move the magnet at different speeds inside the coil, and record the current in the Data and Observations section.

7. Watch the meter, and move the bar magnet in different ways around the outside of the coil. Record your observations.

8. Repeat steps 5 and 6, keeping the magnet stationary and moving the wire coil.

9. Follow your teacher's instructions for proper cleanup.

250 EXPLORE/EXPLAIN Module: Electromagnetic Forces

Data and Observations

Analyze and Conclude

10. What happened to the reading on the ammeter as you moved the magnet?

Generating Electric Current When a magnet is moved through a wire coil that is part of a closed electric circuit, an electric current is produced in the circuit. This happens because the magnetic field from the magnet exerts a force on the charged particles causing them to move in the wire. When the magnet stops moving, there is no current in the circuit. The direction of the current depends on the direction in which the magnet moves. This can be seen in the figure on the right.

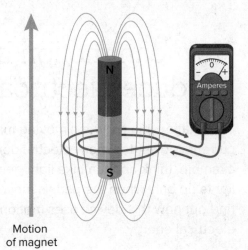

Another way to make an electric current in a wire is to move a wire coil through a magnetic field as shown in the figure below. The magnetic force between the magnet and the charged particles in the wire causes the particles in the wire to move as an electric current in the wire.

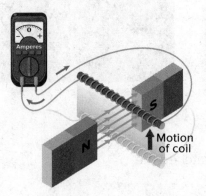
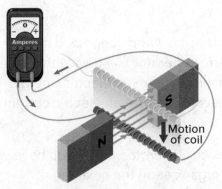

THREE-DIMENSIONAL THINKING

Construct an explanation on the **cause-and-effect** relationship between the reading on the galvanometer or ammeter and the movement of the magnet. Support your explanation with evidence and reasoning.

_____
_____
_____
_____
_____
_____

COLLECT EVIDENCE

How does a magnet produce a current in a wire? Record your evidence (C) in the chart at the beginning of the lesson.

How can motion be used to produce electrical energy?

The relationship between moving magnetic fields and electric currents power the many electric devices you use every day. For example, think about a bike light generator. The light on the bike lights up only when the pedals move. Let's investigate more to find out how this device uses motion and magnets to produce electrical energy.

INVESTIGATION

Lights On

If a bike light does not have a battery it is often connected to a small electric generator. An **electric generator** is a device that uses a magnetic field to transfer mechanical energy to electric energy. How does a generator use magnets to produce an electric current? Let's find out!

GO ONLINE Watch the animation *Generators*. After you have watched the animation, complete the graphic organizer on the next page.

EXPLORE/EXPLAIN Module: Electromagnetic Forces

A wire loop inside the generator is connected in a _____.

The loop is between _____.

As the crank is turned, a _____ rotates through a _____ producing _____ in the circuit.

The current continues as long as the _____ is turned, rotating the _____ within the _____.

Electric Generators In a generator, the crank rotates a wire coil through the magnetic field of a small permanent magnet. This produces an electric current in the circuit. The current continues only as the crank rotates the coil within the magnetic field.

Mechanical Energy to Electric Energy A useful energy transfer occurs in a generator. As you turn a hand generator, it gains mechanical energy. When the wire coils rotate through a magnetic field, an electric current is produced. The rotating coils transfer mechanical energy to electric energy. The more mechanical energy put into the generator, the more electrical energy produced.

Types of Current The electric current produced by a battery differs from the current produced by a generator. The current produced by a battery flows in a circuit in only one direction. An electric current that flows in one direction is **direct current.**

The current produced by a generator changes direction as the poles of the rotating coil line up with the poles of the magnet. An electric current that changes direction in a regular pattern is **alternating current.** You may have seen an alternating current at work during the *Making Magnetic Fields* investigation.

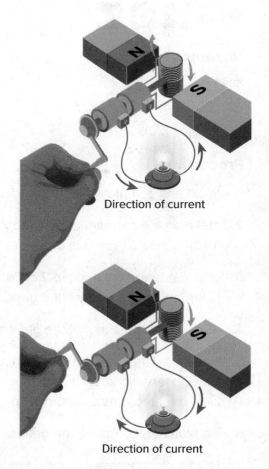

Direction of current

Direction of current

EXPLORE/EXPLAIN Lesson 4 Electromagnetism 253

What affects the strength of the electric current produced by a magnet?

Electric power plants use huge generators, like the ones shown in the image on the right. They generate the electric current used in thousands of homes. How could you increase the strength of the electric current created by a magnet in a generator? Let's find out!

Materials

nail
wire (150 cm)
paper clips (20)
hand generator
sandpaper
compass

Procedure

1. Read and complete a lab safety form.

2. Make an electromagnet by wrapping 130 cm of magnet wire around a nail. Leave 10-cm tails of wire at both ends of the coil.

3. Use sandpaper to remove 2 cm of insulation from each end of the wire. Connect the ends to a hand generator.

4. Slowly turn the crank of the generator.

5. Have a partner use the electromagnet to pick up paper clips. In the Data and Observations section on the next page, record how many paper clips the electromagnet picks up.

6. Repeat step 5, turning the generator at a higher speed.

7. Hold a compass near one end of the coil. Turn the generator at different speeds. Record your observations.

8. Repeat steps 4–7, turning the generator crank in the opposite direction.

9. Follow your teacher's instructions for proper cleanup.

254 EXPLORE/EXPLAIN Module: Electromagnetic Forces

Data and Observations

Rate of Rotation	Number of Paper Clips	Rate of Rotation (opposite direction)	Number of Paper Clips
not rotating		not rotating	
slow		slow	
fast		fast	

Analyze and Conclude

10. Describe the cause-and-effect relationship between the speed of the generator, the direction it was turned, and your results.

_____

_____

_____

_____

11. How does the movement of your hand lead to lifting paper clips?

_____

_____

_____

_____

Electromagnetism All of the phenomena that you have encountered in this lesson are due to electromagnetism. **Electromagnetism** is the interaction between electric charges and magnets. As you have seen, electromagnetism makes many devices work—from MP3 players, to fans, to larger generators used to produce electricity for cities.

Increasing Electric Current Did you notice in the Lab *Lift it up!* that the faster you cranked the generator the more paper clips you were able to pick up? You increased the current in the wire by turning the generator faster, which increased the speed of the magnet. This is just one of the ways you can increase the strength of the current.

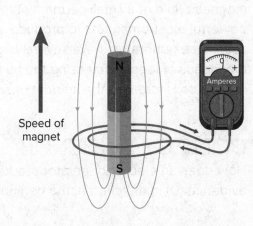

Other ways to increase the strength of the current can be seen in the table below.

The strength of the current is proportional to the strength of the magnet.	The strength of the current increases with the number of coils present in the wire or with the use of an electromagnet or coils around a magnetic core.
Stronger magnet, stronger current 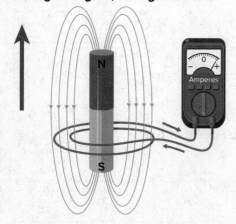	**More coils, stronger current** 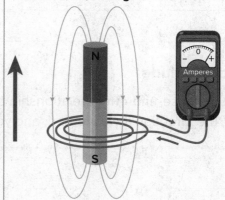
The strength of the current can increase by increasing the speed of the magnet.	The strength of the current can increase by making the magnet more perpendicular to the current.
Faster motion, stronger current 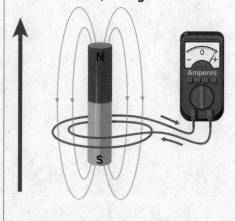	**Less perpendicular, weaker current** 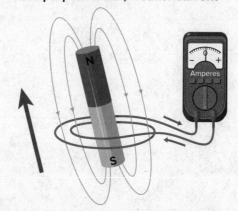

The hand-cranked generator that you used in the Lab *Lift It Up* uses the magnetic field of a small permanent magnet. Larger generators often use powerful electromagnets to produce a magnetic field. Instead of one wire coil, these generators have several large coils of wire. Each coil might have thousands of loops. Increasing the number of coils and the number of loops in each coil increases the amount of current a generator produces.

COLLECT EVIDENCE
How does an electric generator produce electric energy? Record your evidence (D) in the chart at the beginning of the lesson.

256 EXPLORE/EXPLAIN Module: Electromagnetic Forces

STEM Careers

A Day in the Life of a Maglev Train Engineer

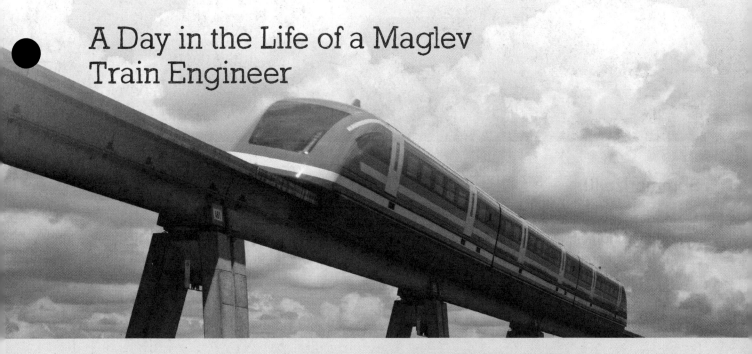

Not all trains run on wheels. Some float! Using the properties of electromagnetic forces, magnetic levitation trains, or "maglev" trains, float along tracks to get from one destination to the other fast. The fastest speed achieved by a maglev train was in Japan at 375 mph!

There are a few ways that a maglev train can achieve suspension in the air. One method is known as electromagnetic suspension. In electromagnetic suspension, strong electromagnets are attached to the train. These are repelled by permanent magnets contained in the track. This causes the train to float above the tracks. Guidance coils keep the train from running into the sides of the track. Engineers must constantly check and update the electromagnets and the magnets' interaction to make sure the train runs smoothly.

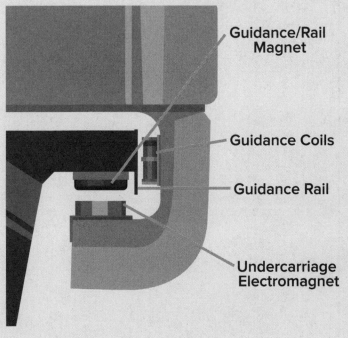

It's Your Turn

Research What other devices use electromagnetic forces to function? Research and create a slideshow presentation on the role electric and magnetic forces play in the device to allow it to work. Use different forms of multimedia in your presentation to support your findings. Share your presentation with your class.

LESSON 4
Review

Summarize It!

1. **Model** Develop a flow chart that shows how an electric motor system is different from an electric generator system. Include the types of energy inputs and outputs, forces, and fields involved in each device.

Three-Dimensional Thinking

Use the figure below to answer questions 2–4.

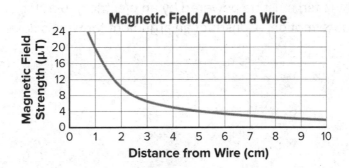

Magnetic Field Around a Wire

2. Estimate the magnetic field strength 3 cm from the wire.

3. Estimate the distance from the wire where the magnetic field strength is 4 μT.

4. How much larger is the magnetic field strength 1 cm from the wire compared to 5 cm from the wire?

   A It is the same.

   B It is twice as large.

   C It is three times as large.

   D It is five times as large.

EVALUATE Lesson 4 Electromagnetism

Real-World Connection

5. Argue Tim found a battery powered hand-held fan. He says that the fan must have an electric motor to power the blades of the fan. Ahisa responds, "No, the blades of the fan must be powered by an electric generator because they produce motion." Construct an argument for who is correct.

_____
_____
_____
_____

6. Explain Seki wants to build a treadmill that is connected to a small generator that would power a phone charger. She is worried that just walking would not create a strong enough current to power a phone for long. Explain the causes and effects for how she could increase the current.

_____
_____
_____
_____

 Still have questions?
Go online to check your understanding about how electric and magnetic forces are related.

REVISIT SCIENCE PROBES
Do you still agree with the statement you chose at the beginning of the lesson? Return to the Science Probe at the beginning of the lesson. Explain why you agree or disagree with that statement now.

PLAN AND DESIGN
STEM Module Project Engineering Challenge
Now that you've learned about the relationships between electric and magnetic forces, go to your Module Project to revise the forces used in your sandbox crane, and start to develop models for your crane. Keep in mind the role that magnetic and electric forces play.

EXPLAIN THE PHENOMENON

Revisit your claim on why the metal detector works only when the current is on. Review the evidence you collected. Explain how your evidence supports your claim.

260 EVALUATE Module: Electromagnetic Forces

STEM Module Project
Engineering Challenge

The Great Metal Pick Up Machine

Your friend's younger brother, Andy, loves his sandbox and all types of construction machinery. His favorite machine is the big crane that can pick up a lot of scrap metal at once. He thinks they are so cool!

Your friend has asked for your help to build Andy his very own crane that he can use in his sandbox to pick up his metal toy cars. First, you will need to determine which type of forces the crane might use. You will then design and test your crane. Using the data from each test, you will optimize your crane design before presenting the crane to your friend.

Planning After Lesson 1

Use the following questions to help you determine a list of the criteria and constraints that will guide your work. Record your list in your Science Notebook.

- Who will use the device?
- What is the device used for?
- What would be evidence that the device is successful?
- What would be evidence that the device is unsuccessful?
- What are reasonable characteristics (for example, size, strength, cost) of the device?
- What must the device be able to lift?

STEM Module Project
Engineering Challenge

Planning After Lesson 2

You have now learned about magnetic forces and electric forces. Construct an argument on which forces might be present in a metal pick-up machine. Include in your argument which forces you would use for the sandbox crane and why.

Planning After Lesson 3

You have now learned about circuits. In the space below, list three possible ways a circuit could be incorporated into the sandbox crane. Chose one of the options and model the needed components of the circuit.

Planning After Lesson 4

You have now learned how electric and magnetic forces interact. Discuss if you want to change the types of forces to incorporate into your sandbox crane. Draw your plans for at least two different design solutions that you will build and test. Label the components of each design. Indicate how the solutions are similar and how they are different.

Develop a list of materials you will need to build and test your device in the space below. Once you receive your teacher's approval, gather the materials and construct your designs.

STEM Module Project: Engineering Challenge

Collect Data

Revisit the criteria that you identified after Lesson 1. Write a procedure to test your design solutions against these criteria. Develop a list of materials you will need to test your device. Your goal is to collect data that you can analyze to identify the best characteristics of each design solution that can be combined into a new solution that better meets the criteria for success.

Conduct the investigation. Organize and record your data and observations in your Science Notebook.

Evaluate your procedure. Did it produce accurate data to show that your design uses noncontact forces through a field? List two ways you could improve your testing procedure.

Analyze Data

Analyze the data collected by your class to determine similarities and differences among design solutions. Based on the analyzed data, identify the best features of each design that can be combined into a new and improved solution.

Select at least two structures in your solution that you want to improve the function of.

Structure 1
What are you trying to improve? How will you improve it?

Structure 2
What are you trying to improve? How will you improve it?

Build a revised design solution using the new features. Repeat your testing procedure. Continue to make modifications in between each test to develop the best design. Record your data and observations in your Science Notebook.

STEM Module Project
Engineering Challenge

Design Your Model

Describe how you used the data generated to improve your design solution.

Construct a model that illustrates how your sandbox crane design uses electromagnetism to pick up a toy car. Include in your system model the potential energy levels, forces, and field present.

Construct an explanation using your model to support the following statement: A force can be exerted on an object even if the objects are not touching causing energy to be transferred to or from the object.

Congratulations! You've completed the Engineering Challenge requirements!

Module Wrap-Up

REVISIT THE PHENOMENON

Think about everything you have learned in the module about magnetic and electric forces. Construct an explanation on how a battery powered fan is related to a hand-cranked flashlight.

OPEN INQUIRY

What are one or two questions you still have about the phenomenon?

Choose the question that interests you the most. Plan and conduct an investigation to answer this question.

EVALUATE Module: Electromagnetic Forces

Glossary

Multilingual Glossary

GO ONLINE to find multilingual glossaries for science. The glossaries include the following languages.

Arabic	Korean	Tagalog
Bengali	Mandarin Chinese	Urdu
French	Portuguese	Vietnamese
Haitian Creole	Russian	
Hmong	Spanish	

Cómo usar el glosario en español:
1. Busca el término en inglés que desees encontrar.
2. El término en español, junto con la definición, se encuentran en la columna de la derecha.

Pronunciation Key

Use the following key to help you sound out words in the glossary.

a	back (BAK)		ew	food (FEWD)
ay	day (DAY)		yoo	pure (PYOOR)
ah	father (FAH thur)		yew	few (FYEW)
ow	flower (FLOW ur)		uh	comma (CAH muh)
ar	car (CAR)		u (+ con)	rub (RUB)
e	less (LES)		sh	shelf (SHELF)
ee	leaf (LEEF)		ch	nature (NAY chur)
ih	trip (TRIHP)		g	gift (GIHFT)
i (i + com + e)	idea (i DEE uh)		j	gem (JEM)
oh	go (GOH)		ing	sing (SING)
aw	soft (SAWFT)		zh	vision (VIH zhun)
or	orbit (OR buht)		k	cake (KAYK)
oy	coin (COYN)		s	seed, cent (SEED)
oo	foot (FOOT)		z	zone, raise (ZOHN)

English — **A** — **Español**

acceleration/direct current | **aceleración/corriente directa**

acceleration: a measure of the change in velocity during a period of time.
alternating current: an electric current that changes direction in a regular pattern.

aceleración: medida del cambio de velocidad durante un periodo de tiempo.
corriente alterna: corriente eléctrica que cambia la dirección en un patrón regular.

— **C** —

closed circuit: circuit is complete and electric energy flows through the circuit.
conduction (kuhn DUK shun): the transfer of charged particles between two conductors.
contact force: a push or a pull on one object by another object that is touching it.

circuito cerrado: circuito que es completo y la energía eléctrica fluye por él.
conducción: movimiento de las partículas cargadas entre dos conductores.
fuerza de contacto: empuje o arrastre ejercido sobre un objeto por otro que lo está tocando.

— **D** —

direct current: an electric current that continually flows in one direction.

corriente directa: corriente eléctrica que fluye de manera continua en una dirección.

displacement: the difference between the initial, or starting, position and the final position.

E

elastic collision: when colliding objects bounce off each other.
elastic potential energy: energy stored in objects that are compressed or stretched, such as springs and rubber bands.
electrically charged: the condition of having an unbalanced amount of positive charge or negative charge.
electrically neutral: a particle with equal amounts of positive charge and negative charge.
electric conductor: a material through which electrons easily move.
electric current: the movement of electrically charged particles.
electric field: the invisible region around a charged object where an electric force is applied.
electric generator: a device that uses a magnetic field to transform mechanical energy to electric energy.
electric insulator: a material through which electrons cannot easily move.
electric motor: a device that uses an electric current to produce motion.
electromagnet: a magnet created by wrapping a current-carrying wire around a ferromagnetic core.
electromagnetism: the interaction between electric charges and magnets.

F

field: a region of space that has a physical quantity (such as a force) at every point.
force: a push or a pull on an object.
force pair: the forces two objects apply to each other.
free-body diagram: a simple model to understand what will happen to an object due to a force.
friction: a contact force that resists the sliding motion of two surfaces that are touching.

G

gravitational potential energy: stored energy due to the interactions of objects in a gravitational field.
gravity: an attractive force that exists between all objects that have mass.

---

desplazamiento: diferencia entre la posición incial y la posición final.

E

choque elástico: ocurre cuando objetos múltiples se chocan.
energía potencial elástica: energía almacenada en objetos que son comprimidos o estresados tal como resortes y banditas elásticas.
cargado eléctricamente: condición de tener una cantidad no balanceada de carga positiva o negativa.
eléctricamente neutro: partícula con cantidades iguales de carga positiva y negativa.
conductor eléctrico: material a través del cual se mueven los electrones con facilidad.
corriente eléctrica: movimiento de partículas cargadas eléctricamente.
campo eléctrico: región invisible alrededor de un objeto cargado en donde se aplica una fuerza eléctrica.
generador eléctrico: aparato que usa un campo magnético para transformar energía mecánica en energía eléctrica.
aislante eléctrico: material por el cual los electrones no pueden fluir con facilidad.
motor eléctrico: aparato que usa corriente eléctrica para producir movimiento.
electroimán: imán fabricado al enrollar un alambre que transporta corriente alrededor de un núcleo ferromagnético.
electromagentismo: interacción entre cargas eléctricas e imanes.

F

campo: región de espacio que tiene una cantidad física (como la fuerza) en todos los puntos.
fuerza: empuje o arrastre ejercido sobre un objeto.
par de fuerzas: fuerzas que dos objetos se aplican entre sí.
diagrama de cuerpo libre: modelo simple para comprender lo que pasará a un objeto debido a la fuerza.
fricción: fuerza de contacto que resiste el movimiento de dos superficies que están en contacto.

G

energía potencia gravitatoria: energía almacenada debida a las interacciones de objetos en un campo gravitacional.
gravedad: fuerza de atracción que existe entre todos los objetos que tienen masa.

I

induction: charging an object without touching it.

K

kinetic (kuh NEH tik) energy: energy due to motion.

L

law of conservation of energy: law that states that energy is always transferring, but energy is not created or destroyed.

M

magnet: an object that attracts iron and other materials that have magnetic qualities similar to iron.
magnetic domain: region in a magnetic material in which the magnetic fields of the atoms all point in the same direction.
magnetic force: a force of attraction or repulsion created by a magnet.
magnetic pole: the place on a magnet where the force it exerts is the strongest.
magnetic potential energy: stored energy due to the interactions of magnetic poles in a magnetic field.
mechanical energy: sum of the potential energy and the kinetic energy in a system.
motion: the process of changing position.

N

net force: the combination of all the forces acting on an object.
Newton's first law of motion: an object at rest will stay at rest, and an object in motion will stay in motion unless a force acts on the object.

Newton's second law of motion: law that states that the acceleration of an object is equal to the net force exerted on the object divided by the object's mass.
Newton's third law of motion: when an object applies a force on another object, the second object applies a force of the same strength on the first object but the force is in the opposite direction.
noncontact force: a force that one object applies to another object without touching it.
normal force: the force that pushes perpendicular to the object's surface.

I

inducción: la carga un objeto por contacto.

K

energía cinética: energía debida al movimiento.

L

ley de la conservación de la energía: ley que mantiene que la energía se puede transformar de una forma a otra, pero no se puede crear ni destruir.

M

imán: objeto que atrae el hierro y otros materiales que tienen calidades magnéticas parecidas al hierro.
dominio magnético: región en un material magnético en el que los campos magnéticos de los átomos apuntan en la misma dirección.
fuerza magnética: fuerza de atracción o repulsión creada por un imán.
polo magnético: lugar en un imán donde la fuerza que éste ejerce es la mayor.
energía potencial magnética: energía almacenada debida a las interacciones de polos magnéticos en un campo magnético.
energía mecánica: suma de la energía potencial y la energía cinética en un sistema.
movimiento: proceso de cambiar de posición.

N

fuerza neta: combinación de todas las fuerzas que actúan sobre un objeto.
primera ley del movimiento de Newton: sostiene que un objeto en reposo permanecerá en reposo y un objeto en movimiento permanecerá en movimiento a menos que una fuerza externa actúe sobre él.
segunda ley del movimiento de Newton: ley que establece que la aceleración de un objeto es igual a la fuerza neta que actúa sobre él divida por su masa.
tercera ley del movimiento de Newton: sostiene que cuando un objeto aplica una fuerza en otro objeto, el segundo objeto aplica una fuerza de la misma intensidad en el primer objeto pero la fuerza está en la dirección opuesta.
fuerza de no contacto: fuerza que un objeto puede aplicar sobre otro sin tocarlo.
fuerza normal: fuerza que empuja perpendicularmente a la superfice de un objeto.

O

open circuit: a circuit that is not complete and no electric energy flows through the circuit.

P

position: an object's distance and direction from a reference point.

potential (puh TEN chul) energy: stored energy due to the interactions between objects or particles.

R

reference point: the starting point you choose to describe the location, or position, of an object.

S

speed: a measure of the distance an object travels in a given amount of time.

V

vector: a quantity that has both magnitude and direction.

velocity: the speed and the direction of a moving object.

voltage: the electrical potential energy difference between two places on an electric circuit.

W

weight: the gravitational force exerted on an object.

work: the transfer of energy to an object by a force that makes an object move in the direction of the force.

O

circuito abierto: circuito que no es completo y no tiene un flujo de energía eléctrica.

P

posición: distancia y dirección de un objeto según un punto de referencia.

energía potencial: energía almacenada debido a las interacciones entre objetos o partículas.

R

punto de referencia: punto que se usa para describir la ubicación o posición de un objeto.

S

rapidez: medida de la distancia que recorre un objeto en un tiempo determinado.

V

vector: cantidad que tiene la magnitud y la dirección.

velocidad: rapidez y dirección de un objeto en movimiento.

voltaje: diferencia del potencial de energía eléctrica entre dos puntos en un circuito eléctrico.

W

peso: fuerza gravitacional ejercida sobre un objeto.

trabajo: transferencia de energía a un objeto por una fuerza que hace que el objeto se mueva en la dirección de la fuerza.

Index

Italic numbers = illustration/photo
Bold numbers = vocabulary term
lab = indicates entry is used in a lab
inv = indicates entry is used in an investigation
smp = indicates entry is used in a STEM Module Project
enc = indicates entry is used in an Encounter the Phenomenon
sc = indicates entry is used in a STEM Career

Acceleration — Friction

A

Acceleration
defined, **40**
force and, 43
gravity and, 88
mass and, 84
Newton's second law and, 44
Air resistance, 155
Alternating current, 253
Ammeter, 252
Amundsen, Roald, 185
Atoms, 191
Attraction, 177

B

Balanced forces, 52, 66.
see also **Force**

C

Cardinal directions, 14
Charged particles
circuits and, 224
in different materials, 210–211
electric forces and, 204
force strength and, 207
magnetic fields and, 240
Circuits
closed v. open, 224
connections and, 222
current and, 225–226 *lab*
electric energy and, 228
Closed circuits, 224, 227, 251
Collision forces, 68–69 *lab*, 70, 71
Compasses, 184
Conduction, 212
Conservation of Energy, 141–160
Contact force, 43. see also **Force**

D

Direct current, 253
Direction
describing, 10–11 *inv*
force and, 43
location and, 11
motion and, 25 *inv*
position and, 11
in two dimensions, 13–14 *inv*
units and, 13–14 *inv*
velocity and, 26
Displacement, 19
Distance
describing, 13–14 *inv*
direction and, 11

displacement and, 19
GPS and, 24
graphs and, 27–28 *inv*, 29
gravity and, 84
location and, 11
motion and, 16
position and, 11
speed and, 22

E

Earth's magnetic field, 185
Elastic collision, 70
Elastic potential energy, 131
Electric charges
positive v. negative, 204, 210–211
potential energy and, 210
transfers of, 211–212
Van de Graaf generators
and, 213
Electric conductors, 211, 224
Electric current
alternating v. direct, 253
circuits and, 225–226 *lab*
defined, **224**
magnetic fields and, 238–239
lab, 240, 241
strength and, 241
Electric devices, 224
Electric energy. see also **Electric potential energy**
circuits and, 224, 227
mechanical energy and, 249, 253
motion and, 252
power grid and, 229
Electric fields, 204, 207, 208–209 *inv*
Electric forces, 197–216
Electric generators
defined, **252**
electromagnets and, 256
magnetic fields and, 253, *254*
mechanical energy and, 253
strength and, 254–255 *lab*, 256
Electric insulators, 211
Electric motors, 248, 249
Electric potential energy, 132
Electric potential energy, 210, 227, 228
Electrically charged, 210
Electrically neutral, 211
Electromagnetic suspension, 249 *sc*
Electromagnetism, 233–260, **255**
Electromagnets
defined, **242**
generators and, *253*

strength and, 246
trains and, 257 *sc*
usefulness of, 243–245
lab, 246
Encounter the Phenomenon, 3, 7, 35, 59, 77, 105, 109, 127, 143, 169, 173, 199, 219, 235
Energy. see **Electric energy**
Energy bars
conservation of energy
and, *150*
energy transfer and, *149*
kinetic energy and, *116*, *120*
potential energy and, *136*
thermal energy and, *155*
work and, *153*
Energy transfer
hydroelectric power and, 157
motion and, 151–152 *lab*,
154–155 *lab*
thermal energy and, 156
work and, 153
Engineers, 137 *sc*
vehicle crash tests and, 53
Entropy, 156
Explain the Phenomenon, 8–9, 32, 36–37, 56, 60–61, 74, 78–79, 94, 103, 110–111, 124, 128–129, 140, 144–145, 160, 167, 174–175, 196, 200–201, 216, 220–221, 232, 236–237, 260, 267

F

Ferromagnetic elements, 177
Fields, 81–82 *inv*
First law of thermodynamics, 156
Force
acceleration and, 41–42 *inv*
defined, **43**
free-body diagrams and, 48
friction and, 47
gravity and, 82–91
magnetic fields and, 184
magnetic poles and, **180,** 181
magnetic potential energy
and, 188
magnets and, 177
Newton's second law and, 44
Force and Acceleration, 33–56
Force pairs, 57–74, **66**
Free-body diagrams, 48, 49, 65–66 *inv*
Friction, 45–46 *lab*, **46,** 47, 155

G

Galvanometers, 252
Global Positioning System (GPS), 24
Graphs
    motion and, 26, 27–28 *inv,* 29
    slope and, 27–28 *inv,* 29
Gravitational acceleration, 87–88
Gravitational Force, 75–94
Gravitational potential energy
    changes in, 135–136
    conservation of energy and, 150
    defined, **135**
    energy transfer and, 149
    explained, 133–135 *inv*
    mass and, 135–136
    other types of potential energy and, 132
    roller coasters and, 137 *sc*
Gravity
    acceleration and, 88
    defined, **82**
    distance and, 84
    fields and, 81–82 *inv*
    force and, 83–84 *inv*
    mass and, 83–84 *inv,* 90 *inv*
    Newton's Principia and, 85
    normal force and, 89
    space travel and, 91
    weight and, 89

H

Hydroelectric power, 157

I

Induction, 211
Inelastic collision, 70
International System of Units (SI), 14
Inverse proportional relationships, 84
Investigation
    Back to Back, 65–66
    Diagram a Force, 48
    Dropping the Ball, 133–135
    Electric Motor Mechanics, 248–249
    Field Rings, 208–209
    Follow the Directions, 10–11
    Graph it, 27–28
    Gravity of Objects, 90
    Lights on, 252–253
    Making Magnetic Fields, 240–241
    Point the Way, 25
    Rolling On, 112
    See You Soon!, 13–14
    Start from Here, 11–12
    The Force of Gravity, 83–84
    The Pencil Dropped Around the World, 81–82
    When Push Comes to Shove, 41–42
Iron, 176–177 *lab*

J

Joules (J), 148

K

Kinetic Energy, 107–124
Kinetic energy
    changes in, 149, 150
    hydroelectric power and, 157
    mass and, 113, 113–115 *lab,* 116
    motion and, **113**
    potential energy and, 146–148 *lab,* 149, 150
    speed and, 117–119 *lab,* 120
    translational v. rotational, 121
    volume and, 113–115 *lab*

L

Lab
    A Balancing Act, 50–51
    Be the Fastest, 20–21
    Bounce Back, 68–69
    Coiled Up, 250–251
    Create a Magnet, 190–191
    Double Pendulum, 154–155
    Electromagnetic Challenge, 243–245
    From Top to Bottom, 202–203
    Lift It Up, 254–255
    Lighten Up, 222–223
    Magnetic Fields, 182–183
    Magnetic Personality, 180–181
    Mass matters, 113–115
    Motor On, 247–248
    Moving Magnets, 187–188
    Paper Clip Pick Up, 176–177
    Paper Pickup, 205–207
    Picking Up Speed, 117–119
    Pointing Directions, 238–239
    Power Up, 225–226
    Pulling Your Weight, 62–63
    Slingshot Physics, 130–131
    So Much Work, 151–152
    Sticky Situation, 45–46
    The Energy of a Pendulum, 146–148
    The Strength of Magnets, 178–179
    Up to Speed, 38–40
    Use the Forces, 80–81
    Watch It Go, 15–17
    Weighing Washers, 87–88
Law of conservation of energy, 150, 151–152 *lab,* 156
Location, GPS and, 24

M

Maglev trains, 257 *sc*
Magnetic domains, 191, 192
Magnetic fields
    atoms and, 191
    bird migration and, 193
    compasses and, 184
    current and, 238–239 *lab,* 240, 241
    Earth and, 185
    electric current generation and, *251*
    energy and, 188
    force and, *184*
    magnetic domains and, 191, 192
    magnets and, 182–183 *lab*
    strength and, 241
Magnetic force
    attraction and repulsion and, 177
    current and, 248–249 *inv*
    magnetic potential energy and, 189
    magnets and, 177
    strength and, 178–179 *lab,* 186
Magnetic Forces, 171–196
Magnetic materials, 191
Magnetic poles
    changeability of Earth's, 185
    force and, **180,** 181
    magnetic fields and, 188
Magnetic potential energy, 132, **188,** 189
Magnetic Resonance Imaging (MRI), 186
Magnets
    closed circuits and, 251
    defined, **177**
    electric motors and, 248
    strength and, 186
    temporary v. permanent, 192
Mass
    gravitational acceleration and, *88*
    gravitational potential energy and, 135–136, *135*
    gravity and, 82, 83–84 *inv,* 90 *inv*
    kinetic energy and, 113, 113–115 *lab,* 116
    Newton's second law and, 44
Mathematical models, 44
Mechanical energy, 253
    air resistance and, 155
    defined, **148**
    energy transfer and, 151–152 *lab*
    potential and kinetic energy and, 146–148 *lab*
Migration, 193
Motion, 81, 252
    acceleration and, 38–40 *lab*
    balanced forces and, 52
    causes of, 112
    changes in, 38–40 *lab*
    defined, **17**
    direction and, 25 *inv*
    distance and, 16
    energy transfer and, 154–155 *lab*
    factors involved in, 15
    kinetic energy and, 113
    Newton's first law of motion, 51

Motion

reference point and, 16
speed and, 22
unbalanced forces and, 52
work and, 153
Motion diagrams, 26

N

Negative charges, 204, 210–211
Net force, 49, 66, 67. *see also* Force
Newitt, Larry, 185
Newton, Sir Isaac, 85. *see also* Newton's first law of motion; Newton's second law of motion; Newton's third law of motion
Newtons (N), weight and, 89
Newton's first law of motion
balanced forces and, 52
defined, **51**
unbalanced forces and, 52
vehicle crash tests and, 53
Newton's Principia, 85
Newton's second law of motion
defined, **44**
friction and, 47
gravitational acceleration and, 88
vehicle crash tests and, 53
Newton's third law of motion
balanced forces and, 66
collision forces and, 68–69 *lab,* 70
defined, **64**
force pairs and, 66
modeling of, 65–66 *inv*
net force and, 67
SAFER and, 71
Noncontact force
defined, **81**
fields and, 81–82 *inv*
gravity and, 81–82 *inv*
motion and, 80–81 *lab*
Nonmagnetic materials, 191
Normal force, 66, 89

O

Open circuits, 224
Opposing forces, 62–63 *lab,* 64

P

Permanent magnets, 192
Position
cardinal directions and, 14
defined, **11**
describing, 10–11 *inv*
direction and, 11
displacement and, 19
distance and, 11
location and, 11–12 *inv*
motion and, 15–17 *lab*
reference point and, 11
Position and Motion, 3–32
Positive charges, 204, 210–211

Potential energy, 125–140, 187–188 *lab,* 210
changes in, 133, 149, 150
defined, **131**
energy change and, 130–131 *lab*
gravity and, 133–135 *inv*
hydroelectric power and, 157
in systems, 132
kinetic energy and, 146–148 *lab,* 149, 150
types of, 132
Power failures, 229
Power grid, 229
Proportional relationships, 116, 120
Proportional relationships, 84

R

Reference direction, 11–12 *inv,* 14
Reference point
defined, **11**
direction and, 13–14 *inv*
GPS and, 24
motion and, 15–17 *lab,* 16
reference direction and, 11–12 *inv*
Repulsion, 177
Review
Lesson 1, 30–32, 122–124, 194–196
Lesson 2, 54–56, 138–140, 214–216
Lesson 3, 72–74, 158–160, 230–232
Lesson 4, 92–94, 258–260
Roller coasters, 137 *sc*
Rotational kinetic energy, 121

S

Satellites, 24
Science Probe
Ball Toss, 75, 94
Blowing in the Wind, 57, 74
Charged Magnets, 233, 260
Constant Mowing, 33, 56
Don't Fall, 125, 140
Electric Charge, 197, 216
Plugging In, 217, 232
Soccer Ball, 107, 124
Swing Low, 141, 160
Train Ride, 5, 32
Which pole is it?, 171, 196
Second law of thermodynamics, 156
Simple Circuits, 217–232
Slope, 27–28 *inv,* 29
Smart meters, 229
Solar system, 86
Space travel, 91
Speed, 117–119 *lab,* 120
average speed equation, 22–23
constant v. changing, 22
defined, **22**
explanation of, 20–21 *lab*
direction and, 25 *inv*

GPS and, 24
slope and, 27–28 *inv,* 29
Springs, 132
Steel and Foam Energy Reduction (SAFER), 71
STEM Careers
A Day in the Life of a Maglev Train Engineer, 257
A Day in the Life of a Roller Coaster Designer, 137
STEM Module Project
Crash Course, 4, 32, 56, 74, 94, 95–102
Energy at the Amusement Park, 106, 124, 140, 160, 161–166
The Great Metal Pick-up Machine, 170, 196, 216, 232, 260, 261–266
Systems, 132

T

Temporary magnets, 192
Thermal energy, 155, 156
Thermodynamics, 156
Time, 27–28 *inv,* 29
Translational kinetic energy, 121

U

Unbalanced forces, 52. *see also* Force
Units, 13–14 *inv*

V

Van De Graaff generators, 213
Vector, 26, 40, 43
Vehicle crash test engineers, 53
Velocity
acceleration and, 40
constant v. changing, **26**
force and, 43
vehicle crash tests and, 53
Voltage, 225–226 *lab,* **227,** 228
Voltmeters, 227
Volume, 113–115 *lab,* 117–119 *lab*

W

Weight, 89
Work, 153